STUDENT'S SOLUTIONS MANUAL TO ACCOMPANY

MULTIPLE-CHOICE & FREE-RESPONSE QUESTIONS IN PREPARATION FOR THE AP CALCULUS (BC) EXAMINATION

(SEVENTH EDITION)

By David Lederman

With the assistance of:

Lin McMullin

D&S MARKETING SYSTEMS, INC.
1205 38th Street • Brooklyn, NY 11218

www.dsmarketing.com

ISBN # 1-878621-99-8

Copyright © 2005 by D & S Marketing Systems, Inc.

All rights reserved.

No part of this book may be reproduced or transmitted in any form or by any means, electronic or mechanical, including photocopying and recording, or by any information storage or retrieval systems, without written permission from the publisher.

Printed in the U.S.A.

PREFACE

This Student's Solutions Manual is a supplement to MULTIPLE-CHOICE & FREE-RESPONSE QUESTIONS IN PREPARATION FOR THE AP CALCULUS (BC) EXAMINATION, SEVENTH EDITION.

This manual is a valuable aid for those who desire to understand the steps involved in working out the problems. You should not rely solely on this manual. Only use it after you have first tried a problem and are "stuck".

I will be delighted to hear from students and teachers who come up with creative or alternate ways of solving the given problems. I would also appreciate hearing from you if you find any errors, typographical or mathematical.

I wish to thank Bob Byrne of St. Thomas Aquinas HS (Fort Lauderdale, FL) for sharing ideas and making valuable suggestions which have been incorporated in this edition. A special thanks to Lin McMullin for contributing many graphing calculator, conceptual, and free-response questions which have been included in this edition. A special debt of gratitude is due to my wife for her encouragement and patience. I also wish to thank Tammi Pruzansky of D&S Marketing Systems, Inc., for preparing the graphs and typesetting the entire manuscript.

Any errors found in this book are solely the responsibility of the author.

All communications concerning this book should be addressed to:

D & S Marketing Systems, Inc.
1205 38th Street
Brooklyn, NY 11218
www.dsmarketing.com

TABLE OF CONTENTS

SAMPLE EXAM I..1

SAMPLE EXAM II..21

SAMPLE EXAM III...39

SAMPLE EXAM IV...59

SAMPLE EXAM V..81

SAMPLE EXAM VI...100

Sample Examination I

1. To be continuous both parts of the function must have the same value at $x = 5$.

$$5^2 + 5b = 5\sin\left(\frac{\pi}{2}(5)\right)$$
$$25 + 5b = 5(1)$$
$$5b = -20$$
$$b = -4$$

The correct choice is (C).

2. $y = 3x^2 - x^3$

$y' = 6x - 3x^2 = 3x(2-x)$

The critical points are at $x = 0$ and $x = 2$.

Using the First Derivative Test to find possible relative max/min,

	$x < 0$	$0 < x < 2$	$x > 2$
y'	$-$	$+$	$-$

The only relative maximum occurs at $x = 2$ (where y' switches from $+$ to $-$).

Therefore, $(2, 4)$ is the only relative maximum point of $y = 3x^2 - x^3$.

The correct choice is (C).

3. $v(t) = (t^2, \sin(\pi t))$

$$s(t) = \left(\frac{t^3}{3} + C, -\frac{1}{\pi}\cos(\pi t) + K\right)$$

Since $s(0) = (1, 0)$ from the given information,

$$s(t) = \left(\frac{t^3}{3} + 1, -\frac{1}{\pi}\cos(\pi t) + \frac{1}{\pi}\right), \text{ and}$$

$$s(3) = \left(10, -\frac{1}{\pi}\cos(3\pi) + \frac{1}{\pi}\right) = \left(10, \frac{2}{\pi}\right)$$

The correct choice is (B).

4. Use implicit differentiation to find $\dfrac{dy}{dx}$:

$$2y\frac{dy}{dx} - 3x^2 - 30x = 0$$
$$2y\frac{dy}{dx} = 3x^2 + 30x$$
$$\frac{dy}{dx} = \frac{3x(x+10)}{2y}$$

The derivative is zero when $x = 0$ and $x = -10$.

The derivative may also be found by solving for y and then differentiating:

$$y = \pm\sqrt{x^3 + 15x^2 + 8}$$
$$\frac{dy}{dx} = \frac{3x^2 + 30x}{\pm 2\sqrt{x^3 + 15x^2 + 8}}$$
$$= \frac{3x(x+10)}{\pm 2\sqrt{x^3 + 15x^2 + 8}}$$

As above the derivative is zero when $x = 0$ and $x = -10$.

The correct choice is (D).

5. As $x \to \infty$, the denominator approaches ∞ faster than the numerator, since $x^6 > x^5$.

Therefore, the entire expression approaches 0 as $x \to \infty$.

The correct choice is (A).

6. The given expression is of the form $\dfrac{0}{0}$, so use L'Hôpital's Rule to find the required limit. The derivative of the integral is found by using the Fundamental Theorem of Calculus.

$$\lim_{x \to 0} \frac{\int_1^{1+x} \frac{\cos t}{t}\,dt}{x} = \lim_{x \to 0} \frac{\frac{\cos(1+x)}{(1+x)}}{1} = \cos 1$$

The correct choice is (B).

7. $f(x) = \sqrt{4\sin x + 2} = (4\sin x + 2)^{\frac{1}{2}}$

$f'(x) = \dfrac{1}{2}(4\sin x + 2)^{-\frac{1}{2}}(4\cos x) = \dfrac{2\cos x}{\sqrt{4\sin x + 2}}$

Therefore, $f'(0) = \dfrac{2\cos(0)}{\sqrt{4\sin(0) + 2}} = \dfrac{2(1)}{\sqrt{0 + 2}} = \dfrac{2}{\sqrt{2}}$ or $\sqrt{2}$.

The correct choice is (E).

8. By the Fundamental Theorem of Calculus $\int_0^2 f'(t)\,dt = f(2) - f(0)$

    Since $f'(x)$ is in knots or nautical miles per hour and t (and therefore dt) is in hours, their product $f'(t)\,dt$ is $\dfrac{\text{nautical miles}}{\text{hour}} \cdot \text{hour} = \text{nautical miles}$.

    $\int_0^2 f'(t)\,dx = f(2) - f(0)$ nautical miles.

    The correct choice is (D).

9. By implicit differentiation of $x^2 + y^2 = 169$, $2x + 2y\dfrac{dy}{dx} = 0$

    Solving for $\dfrac{dy}{dx}$, $\dfrac{dy}{dx} = \dfrac{-x}{y}\bigg|_{(5,-12)} = \dfrac{5}{12}$

    A tangent line to the circle $x^2 + y^2 = 169$ at $(5, -12)$ has a slope of $\dfrac{5}{12}$.

    An equation of the tangent line is $y = y_1 + m(x - x_1)$, or

    $$y = -12 + \dfrac{5}{12}(x - 5)$$

    Multiplying both sides of the equation by 12 and simplifying, $5x - 12y = 169$.

    The correct choice is (C).

10. The speed is the absolute value of the velocity. The speed is the greatest when the velocity is farthest from zero (above or below). This occurs at point D when the (negative) velocity has a larger absolute value than B where the velocity is greatest.

    The correct choice is (D).

11. Applying the ratio test to the given series,

$$\lim_{n\to\infty}\left|\frac{(-1)^{n+1}x^{n+1}}{\ln(n+1)}\div\frac{(-1)^n x^n}{\ln n}\right|=\lim_{n\to\infty}\frac{\ln(n)}{\ln(n+1)}|x|=|x|<1.$$

Therefore, the series converges absolutely if $|x|<1$. Now test the endpoints.

If $x=1$, the series $\sum_{n=2}^{\infty}\frac{(-1)^n}{\ln n}$ converges by the Alternating Series Test.

If $x=-1$, the series $\sum_{n=2}^{\infty}\left(\frac{1}{\ln n}\right)$ diverges by the comparison test (with the harmonic series,

i.e. $\sum_{n=2}^{\infty}\frac{1}{\ln n}>\sum_{n=2}^{\infty}\frac{1}{n}$, which diverges).

Therefore, the interval of convergence is $-1<x\leq 1$.

The correct choice is (D).

12. Recall that $\int\frac{du}{\sqrt{a^2-u^2}}=\text{Arcsin}\left(\frac{u}{a}\right)+C$. Let $u=x$, $du=dx$ and $a=2$.

$$\int\frac{dx}{\sqrt{4-x^2}}=\text{Arcsin}\left(\frac{x}{2}\right)+C$$

Therefore, the correct choice is (A).

13. In general, a function $f(x)$ has a point of inflection where $f''(x)=0$ or when $f''(x)$ is undefined, and the concavity switches at that point.

In the given problem, $f'(x)=4x-\frac{k}{x^2}$ and $f''(x)=4+\frac{2k}{x^3}$.

Since $f(x)$ has a point of inflection at $x=-1$, $4+\frac{2k}{(-1)^3}=4-2k=0$

or $k=2$.

The correct choice is (A).

14. Using integration by parts, $u = x \quad dv = \sec^2 x \, dx$
$$du = dx \quad v = \tan x$$

$$\int x \sec^2 x \, dx = x \tan x - \int \tan x \, dx = x \tan x - (-\ln|\cos x|) + C$$
$$= x \tan x + \ln|\cos x| + C$$

Therefore, $f(x) = x \tan x$.

The correct choice is (C).

15. $\dfrac{dx}{dt} = -3e^{-t}$ and $\dfrac{dy}{dt} = 6e^t$

$$\dfrac{dy}{dx} = \dfrac{\frac{dy}{dt}}{\frac{dx}{dt}} = \dfrac{6e^t}{-3e^{-t}}\bigg|_{t=0} = \dfrac{6}{-3} \text{ or } -2$$

When $t = 0$, $x = 3e^0$ or $x = 3$, and $y = 6e^0$ or $y = 6$.

The equation of the tangent line is $y - 6 = -2(x - 3) = -2x + 6$ or $2x + y - 12 = 0$

The correct choice is (A).

16. By partial fractions,

$$\dfrac{1}{2x^2 + 3x + 1} = \dfrac{A}{2x + 1} + \dfrac{B}{x + 1}$$
$$1 = A(x + 1) + B(2x + 1)$$

If $x = -1$, $1 = -B \Rightarrow B = -1$.

If $x = -\dfrac{1}{2}$, $1 = \dfrac{1}{2}A \Rightarrow A = 2$.

Therefore, $\dfrac{1}{2x^2 + 3x + 1} = \dfrac{2}{2x + 1} - \dfrac{1}{x + 1}$

$$\int \dfrac{dx}{2x^2 + 3x + 1} = \int \dfrac{2}{2x + 1} dx - \int \dfrac{dx}{x + 1} = \ln|2x + 1| - \ln|x + 1| + C = \ln\left|\dfrac{2x + 1}{x + 1}\right| + C$$

The correct choice is (D).

17. $\frac{dx}{dt} = \cos t$ and $\frac{dy}{dt} = -2\sin t \cos t$

$$\frac{dy}{dx} = \frac{\frac{dy}{dt}}{\frac{dx}{dt}} = \frac{-2\sin t \cos t}{\cos t} = -2\sin t$$

$$\frac{d^2y}{dx^2} = \frac{\frac{d}{dt}\left(\frac{dy}{dx}\right)}{\frac{dx}{dt}} = \frac{-2\cos t}{\cos t} = -2$$

Therefore, $\frac{d^2y}{dx^2}$ is always -2.

The correct choice is (A).

18. $y = x(\ln x)^2$

$\frac{dy}{dx} = x(2\ln x)(\frac{1}{x}) + (\ln x)^2 = 2\ln x + (\ln x)^2 = (\ln x)(2 + \ln x)$

The correct choice is (C).

19. Referring to the accompanying diagram,

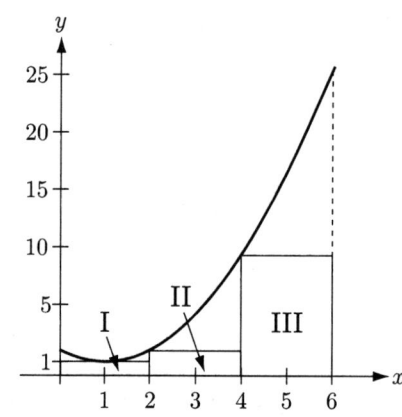

Area of Rectangle I: $2 \cdot f(1) = 2 \cdot 1 = 2$

Area of Rectangle II: $2 \cdot f(2) = 2 \cdot 2 = 4$

Area of Rectangle III: $2 \cdot f(4) = 2 \cdot 10 = 20$

<u>Total</u> Area of Rectangles: $2 + 4 + 20 = 26$

The correct choice is (B).

20. Notice that the segments that make up the slope field are always parallel across the page. This means that the x-coordinates do not affect the slope and therefore x does not appear in the differential equation. This eliminates choices (A), (B) and (C). The slope field for choice (D) can contain only segments with positive slopes. This eliminates (D).

    The correct choice is (E).

21. The series is geometric with a common ratio of $2x$. It will converge when
$$|2x| < 1$$
$$-1 < 2x < 1$$
$$-\tfrac{1}{2} < x < \tfrac{1}{2}$$

    The correct choice is (B).

22. The well known series for $\sin(x)$ is $x - \dfrac{x^3}{3!} + \dfrac{x^5}{5!} - \dfrac{x^7}{7!} + \cdots$

    Substitute x^2 for x to obtain the series for
$$\sin(x^2) = x^2 - \frac{(x^2)^3}{3!} + \frac{(x^2)^5}{5!} - \frac{(x^2)^7}{7!} + \cdots = x^2 - \frac{x^6}{3!} + \frac{x^{10}}{5!} - \frac{x^{14}}{7!} + \cdots$$

    Then divide each term by x^2 to obtain the required series:
$$\frac{\sin(x^2)}{x^2} = 1 - \frac{x^4}{3!} + \frac{x^8}{5!} - \frac{x^{12}}{7!} + \cdots = \sum_{k=0}^{\infty} \frac{(-1)^k x^{4k}}{(2k+1)!}$$

    You may also work directly from the summation notation:
$$\sin(x) = \sum_{k=0}^{\infty} \frac{(-1)^k x^{2k+1}}{(2k+1)!}$$

$$\sin(x^2) = \sum_{k=0}^{\infty} \frac{(-1)^k (x^2)^{2k+1}}{(2k+1)!} = \sum_{k=0}^{\infty} \frac{(-1)^k x^{4k+2}}{(2k+1)!}$$

$$\frac{\sin(x^2)}{x^2} = \frac{1}{x^2} \sum_{k=0}^{\infty} \frac{(-1)^k x^{4k+2}}{(2k+1)!} = \sum_{k=0}^{\infty} \frac{(-1)^k x^{4k}}{(2k+1)!}$$

    The correct choice is (D).

23. The length of the curve $y = f(x)$ from $x = a$ to $x = b$ is $L = \int_a^b \sqrt{1 + [f'(x)]^2}\ dx$.

Therefore, $1 + [f'(x)]^2 = e^{2x} + 2e^x + 2$
$[f'(x)]^2 = e^{2x} + 2e^x + 1 = (e^x + 1)^2$
$f'(x) = e^x + 1$

To find $f(x)$, integrate $f'(x)$: $f(x) = \int (e^x + 1)\ dx = e^x + x + C$.

So any function of the form $f(x) = e^x + x + C$, where C is any constant, will have a length of $\int_a^b \sqrt{e^{2x} + 2e^x + 2}\ dx$.

The only choice of this form is $e^x + x - 2$.

The correct choice is (E).

24. By the Product Rule $h'(x) = f(x)g'(x) + g(x)f'(x)$ and

$h'(3) = f(3)g'(3) + g(3)f'(3) = (1)(1) + (3)(-\frac{1}{3}) = 0$.

The values of the functions are read from the graph; the values of the derivatives are the slopes of the lines at $x = 3$.

The correct choice is (C).

25. See Dominance and Comparisons of Rates of Change on page 216 - 217. If $\lim\limits_{x \to \infty} \dfrac{g(x)}{f(x)} = 0$, then $g(x)$ grows faster than $f(x)$.

You are looking for functions $f(x)$ and $g(x)$ so that $g(x)$ grows faster than $f(x)$ as $x \to \infty$. The only pair of functions is (C), $g(x) = e^x$ grows faster than $f(x) = \ln x$ as $x \to \infty$.

The correct choice is (C).

26. The integral $\int_1^4 \dfrac{t-2}{(t+1)(t-4)}\ dt$ is an improper integral, since the integrand has a vertical asymptote at $t = 4$. Since the limits of integration are $t = 1$ and $t = 4$, you must rewrite $\int_1^4 \dfrac{t-2}{(t+1)(t-4)}\ dt$ as $\lim\limits_{b \to 4^-} \int_1^b \dfrac{t-2}{(t+1)(t-4)}\ dt$.

The correct choice is (E).

27. The average value is $\dfrac{1}{0-(-4)}\displaystyle\int_{-4}^{0}\cos\left(\tfrac{1}{2}x\right)dx = \tfrac{1}{4}\displaystyle\int_{-4}^{0}\cos\left(\tfrac{1}{2}x\right)dx$

Let $u = \tfrac{1}{2}x$, $du = \tfrac{1}{2}dx \Rightarrow dx = 2du$

$$\int \cos\left(\tfrac{1}{2}x\right)dx = 2\int \cos(u)\,du = 2\sin\left(\tfrac{1}{2}x\right)$$

So, $\tfrac{1}{4}\displaystyle\int_{-4}^{0}\cos\left(\tfrac{1}{2}x\right)dx = \tfrac{1}{4}\cdot 2\sin(\tfrac{1}{2}x)\Big|_{-4}^{0} = \tfrac{1}{2}\sin(\tfrac{1}{2}x)\Big|_{-4}^{0} = \tfrac{1}{2}(0-\sin(-2)) = -\tfrac{1}{2}\sin(-2)$.

Since $\sin(-2) = -\sin(2)$, $-\tfrac{1}{2}\sin(-2) = -\tfrac{1}{2}(-\sin(2)) = \tfrac{1}{2}\sin(2)$.

The correct choice is (E).

28. The expression $\displaystyle\lim_{n\to\infty}\dfrac{1}{n}\left[\dfrac{1}{1+(1/n)} + \dfrac{1}{1+(2/n)} + \cdots + \dfrac{1}{1+(n/n)}\right] = \displaystyle\lim_{n\to\infty}\sum_{k=1}^{n} f(x_k)\Delta x$ where $\Delta x = \dfrac{1}{n}$, $x_k = 1 + \dfrac{k}{n}$ and $f(x_k) = \dfrac{1}{x_k}$. The given limit is the right-hand Riemann sum for $y = \dfrac{1}{x}$ from $x=1$ to $x=2$. It is equivalent to $\displaystyle\int_{1}^{2}\dfrac{1}{x}\,dx$.

Note: Another equivalent answer, not among the choices, is $\displaystyle\int_{0}^{1}\dfrac{1}{1+x}\,dx$.

The correct choice is (A).

29. $V_x = \pi\displaystyle\int_{0}^{3}\left[(x-3)^2\right]^2 dx = \pi\displaystyle\int_{0}^{3}(x-3)^4\,dx$

The correct choice is (B).

30. The tangent lines will be parallel when the derivatives of f and g are equal. Solve the equation on your calculator.

$$f'(x) = g'(x)$$
$$\sec^2 x = 2x$$
$$\dfrac{1}{\cos^2(x)} = 2x$$
$$x \approx 2.083$$

The correct choice is (C).

31. The average rate of change of f is given by $\frac{\Delta f}{\Delta t} = \frac{\frac{1}{b} - \frac{1}{a}}{b-a} = \frac{a-b}{ab(b-a)} = -\frac{1}{ab}$. The derivative is $f'(t) = -\frac{1}{t^2}$. Set these equal and solve for t:

$$-\frac{1}{ab} = -\frac{1}{t^2}$$
$$t^2 = ab$$
$$t = \sqrt{ab}$$

The correct choice is (B).

32. $f'(x) = 0 + \cos(x^2) = 0$

$$x^2 = \frac{\pi}{2} + k\pi$$
$$x = \sqrt{\frac{\pi}{2}} \text{ is the smallest positive value}$$
$$x = 1.253$$

The equation may also be solved using a graphing calculator.

The correct choice is (B).

33. By the Fundamental Theorem of Calculus, $\int_0^3 R(t)\, dt$ is the total amount of water that leaks out in the first three hours. Keep in mind that the definite integral of a rate of change gives the total amount of change.

The correct choice is (C).

34.

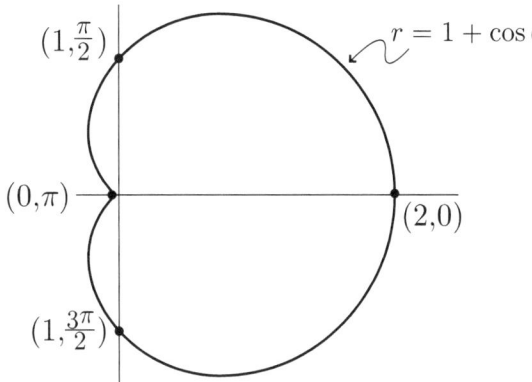

The polar curve $r = 1 + \cos\theta$ is a cardioid which is symmetric about the x-axis as shown in the figure above.

The area is $A = \dfrac{1}{2} \displaystyle\int_0^{2\pi} (1 + \cos\theta)^2 \, d\theta$.

This is <u>not</u> one of the choices; however, since $r = 1 + \cos\theta$ is symmetric about the x-axis, the area can be found by doubling the area above the x-axis, and can be written as

$$2\left(\dfrac{1}{2}\int_0^\pi (1 + \cos\theta)^2 \, d\theta\right) \text{ or the area } A = \int_0^\pi (1 + \cos\theta)^2 \, d\theta.$$

The correct choice is (B).

35. The coordinates of B are $(1000, 100)$. Therefore, substituting into the first part of the equation of R: $100 = a \cdot 1000^2 \Rightarrow a = 10^{-4}$.

The slope of the parabola is $R'(x) = 2ax = 2 \cdot 10^{-4} x$. Since the derivative is continuous, the slope of the line must be the same as the slope of the parabola at B, therefore $b = 2 \cdot 10^{-4} \cdot 1000 = 0.2$.

The correct choice is (A).

36. Since $\sin\left(\dfrac{\pi}{6}\right) = \dfrac{1}{2}$, the series is geometric with a first term of 1 and a ratio of $\dfrac{1}{2}$.

Therefore, the sum is $\dfrac{a_0}{1 - r} = \dfrac{1}{1 - \frac{1}{2}} = 2$.

The correct choice is (B).

37. Examine each of the functions $f(x)$, $g(x)$, and $h(x)$.

   $f(x)$: As x approaches zero, the values appear to be getting closer to 2.

   $g(x)$: Even though $g(0) = 2$, the values to the left of $x = 0$ do not approach 2, so 2 is not the limit.

   $h(x)$: The values of h are approaching 2 as x approaches zero. The fact that the function is not defined at $x = 0$ does not affect the limit.

   The correct choice is (D).

38. Solve the differential equation by separating the variables:

    $$y\, dy = x^2 dx$$

    Antidifferentiate both sides:

    $$\frac{1}{2}y^2 = \frac{1}{3}x^3 + C$$

    Substitute the initial condition ($x = 1$ when $y = 1$) to find the constant C:

    $$\frac{1}{2}(1) = \frac{1}{3}(1) + C \Rightarrow C = \frac{1}{6}$$

    $$\frac{1}{2}y^2 = \frac{1}{3}x^3 + \frac{1}{6}$$

    $$y^2 = \frac{2}{3}x^3 + \frac{1}{3}$$

    $$y = \sqrt{\frac{2x^3 + 1}{3}}$$

    The correct choice is (E).

39. The local maximums and minimums of the derivative correspond to points of inflection of the function. For example, at the top left maximum point the derivative changes from increasing faster to increasing slower, thus the function changes from concave up to concave down. Another way to tell this is that at the extreme points the derivative's derivative, the second derivative, changes sign, indicating a point of inflection. There are four such points shown.

    The correct choice is (E).

40. The average value of a function on an interval $[a, b]$ is given by $\dfrac{1}{b-a}\displaystyle\int_a^b f(x)\,dx$. The integral gives the area under the graph. So in this problem the average value is
$$\dfrac{area}{12} = \dfrac{4 \cdot 8 + \frac{1}{4}\pi(4^2)}{12} \approx 3.714$$

The correct choice is (B).

41. The third-degree Maclaurin polynomial for the $\sin(x)$ is $x - \dfrac{x^3}{3!}$ (this is one of the power series that should be memorized). Graph this and $y = x^2$ in the same window (see Figure 1). The graph shows 2 points of intersection, BUT there is another point. The graph is a cubic function and a quadratic function. <u>Cubic functions grow faster than quadratics</u>. There is a third point of intersection in the second quadrant as figure 2 shows. It should *not* be necessary to produce the second graph to know that they *will* intersect again.

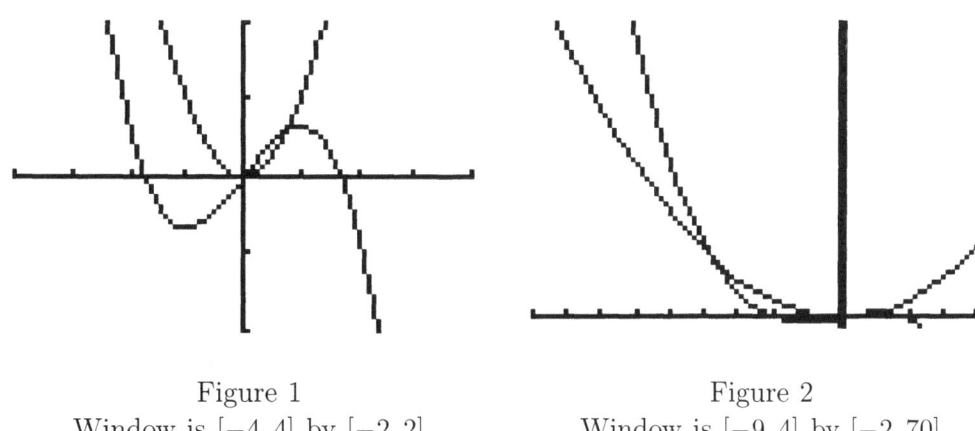

Figure 1
Window is $[-4, 4]$ by $[-2, 2]$

Figure 2
Window is $[-9, 4]$ by $[-2, 70]$

The correct choice is (D).

42. The rate of production is given by the first derivative of A: $A'(t) = 48 - 12(t-3)^2$.

The maximum increase in the rate of production occurs when $A'(t)$ has its maximum value, that is, when $A''(t) = 0$.

$A''(t) = -24(t-3) \Rightarrow A''(t) = 0$ when $t = 3$. Three hours after 8:00 a.m. is 11:00 a.m.

<u>Note</u>: $A'''(3) = -24$ indicating that $t = 3$ is a maximum (Second Derivative Test). Or use the First Derivative Test: A'' switches from positive to negative at $t = 3$, so $t = 3$ is a maximum.

The correct choice is (C).

43. Using the Product Rule the derivative is

$$4x^2(-\sin(x)) + (\cos(x))(8x) = 8x\cos(x) - 4x^2\sin(x).$$

The correct choice is (B).

44. The limiting value is 6000 as time goes on, i.e. $\lim_{t \to +\infty}(6000 - 5500e^{-0.159t}) = 6000$. You want $P(t) = 3000$. This can be easily found by use of a graphing calculator. A pencil and paper solution is also possible.

$$3000 = 6000 - 5500e^{-0.159t}$$

$$-3000 = -5500e^{-0.159t}$$

$$\frac{30}{55} = e^{-0.159t}$$

$$\ln\left(\frac{30}{55}\right) = -0.159t \Rightarrow t = \frac{\ln(\frac{30}{55})}{-0.159} \Rightarrow t \approx 3.8$$

This occurs during the fourth year.

The correct choice is (C).

45. Since $f'(x) > 0$ for $0 < x < 6$ the maximum point on the graph of $f(x)$ is at $x = 6$ the end point.

Therefore, the maximum value of $f(x)$ is $f(6)$. To find $f(6)$, use the Fundamental Theorem of Calculus:

$$\int_2^6 f'(x)\,dx = f(6) - f(2) \text{ or, } f(6) = f(2) + \int_2^6 f'(x)\,dx$$

Since you are given that $f(2) = 10$, $f(6) = 10 + \int_2^6 f'(x)\,dx$.

Now, $\int_2^6 f'(x)\,dx$ is the area under $f'(x)$ from $x = 2$ to $x = 6$, which is approximately 100. (Each square box has an area of 5, and there are approximately 20 square boxes combined.)

Therefore, the maximum value of $f(x) \approx 10 + 100$ or 110.

The correct choice is (E).

1a. At any point the slope of the hill is given by $\frac{d}{dx}f(x) = -\frac{50}{100}\sin\left(\frac{x}{100}\right)$. An equation of the tangent line at $x = a$ is:

$$y - 50\cos\left(\frac{a}{100}\right) = -\frac{50}{100}\left(\sin\left(\frac{a}{100}\right)\right)(x - a)$$

$$y = -\frac{1}{2}\left(\sin\left(\frac{a}{100}\right)\right)x + \frac{a}{2}\sin\left(\frac{a}{100}\right) + 50\cos\left(\frac{a}{100}\right)$$

1b. The top of the hill is at $(0, 50)$ and the person's eyes are at $(0, 55)$, which is the y-intercept of the line. Therefore,

$$\frac{a}{2}\sin\left(\frac{a}{100}\right) + 50\cos\left(\frac{a}{100}\right) = 55$$

Solve this equation on your calculator for values in the approximate interval $[0, 100\pi]$ covering the top to the bottom of the hill. $\underline{a = 45.935}$. (The tangent also intersects the hill at $a = 398.341$, but this is obviously past the middle of the valley.)

1c. The middle of the valley is at $x = 100\pi$ (half the period of $f(x) = 50\cos\left(\frac{x}{100}\right)$). $f(100\pi) = -50$ while $y(100\pi) = -\frac{1}{2}\left(\sin\left(\frac{45.935}{100}\right)\right)(100\pi) + 55 = -14.643$. Since $25 < -14.643 - (-50)$ the top of the house is at $(100\pi - 25)$, which is below the person's line of sight along the tangent line. Hence the tope of the house cannot be seen by the person.

2a.

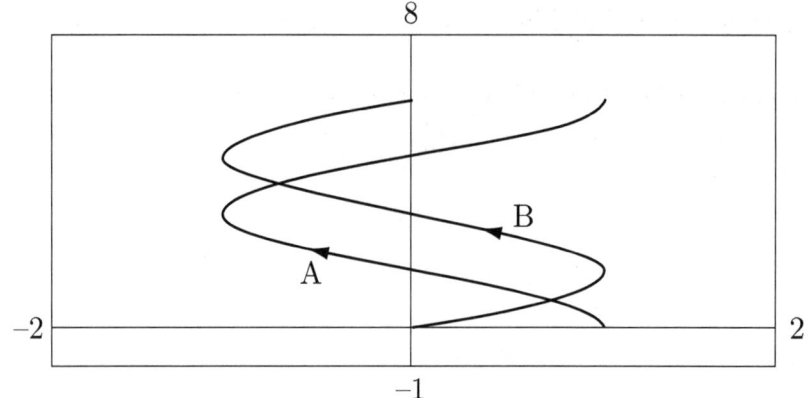

2b. The velocity vector is the derivative of the position vector:

For A: $(-\sin t, 1)$

For B: $(\cos t, 1)$

2c. At $t = 5$, the x-component of the velocity vector for A is $-\sin(5) = 0.959$ and for B it is $\cos(5) = 0.284$. Therefore, A is moving to the right faster.

2d. For all t, including $t = 5$, the y-component of the velocity vector of both particles is the same (i.e. they are both 1) – therefore, both particles are moving upwards at the same rate.

3a. $L'(d) = (167.5)\left(\dfrac{2\pi}{366}\right)\cos\left(\left(\dfrac{2\pi}{366}\right)(d-80)\right).$

The maximum occurs when the derivative changes sign from positive to negative.

$\cos\left(\left(\dfrac{2\pi}{366}\right)(d-80)\right) = 0.$

$\left(\dfrac{2\pi}{366}\right)(d-80) = \dfrac{\pi}{2}$ or $\dfrac{3\pi}{2}$, which means $d = \dfrac{\pi}{2}\dfrac{366}{2\pi} + 80 = 171.5$

or $d = \dfrac{3\pi}{2}\dfrac{366}{2\pi} + 80 = 354.5$. Since $L'(d) > 0$ for $0 < d < 171.5$, and $L'(d) < 0$ for $171.5 < d < 354.5$, the maximum occurs on the 172nd day. (Day 355 is in December and is the minimum.)

3b. The mean value is given by $\dfrac{1}{366}\displaystyle\int_0^{366} L(d)\,dd = 731$, by use of a calculator.

OR: $L(d)$ is a sine function which in each full period has the same area above and below the horizontal line $y = 731$, therefore its average value is 731. The graph of $L(d)$ and $y = 731$ are shown below in the window $x[0, 366]$ by $y[500, 1000]$

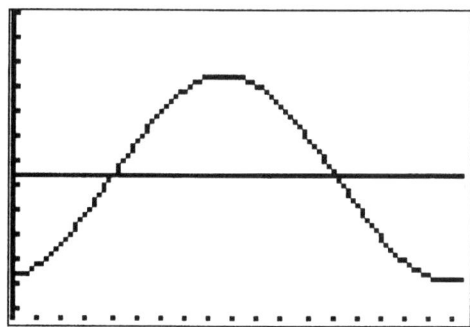

3c. The total number of minutes of daylight is given by $\int_0^{366} L(d)\,dd = 267,546$ minutes. This may also be found by multiplying the average minutes of daylight (731) by the number of days (366).

4a. $\displaystyle\int_1^\infty \frac{1}{x^2}\,dx = \lim_{b\to\infty}\int_1^b \frac{1}{x^2}\,dx = \lim_{b\to\infty}\left(-\frac{1}{x}\right)\Big|_1^b = \lim_{b\to\infty}\left(-\frac{1}{b}-(-1)\right) = 1$

The integral converges to 1.

4b. $\displaystyle\int_1^\infty \left(1-\frac{1}{x^2}\right)\,dx = \lim_{b\to\infty}\int_1^b \left(1-\frac{1}{x^2}\right)\,dx = \lim_{b\to\infty}\left(x+\frac{1}{x}\right)\Big|_1^b = \lim_{b\to\infty}\left(\left(b+\frac{1}{b}\right)-(1+1)\right) = \infty$

The integral diverges.

4c. $\displaystyle\int_1^\infty \left(\frac{1}{x^2}\right)^2\,dx = \lim_{b\to\infty}\int_1^b \frac{1}{x^4}\,dx = \lim_{b\to\infty}\left(-\frac{1}{3x^3}\right)\Big|_1^b = \lim_{b\to\infty}\left(-\frac{1}{3b^3}-\left(-\frac{1}{3}\right)\right) = \frac{1}{3}$

The integral converges to $\frac{1}{3}$.

5a.

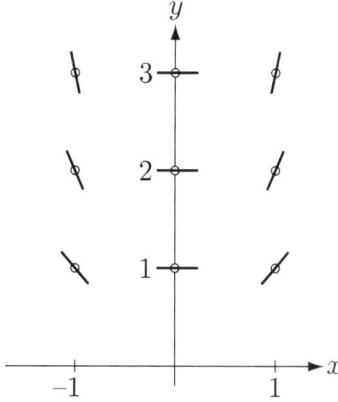

5b. Solve the differential equation by separating the variables and integrating both sides.

$$y^{-2} dy = x \, dx$$
$$\frac{-1}{y} = \frac{x^2}{2} + C$$
$$y = \frac{-1}{\frac{x^2}{2} + C}$$

5c. Substitute the initial condition:

$$\frac{-1}{0^2 + C} = 1$$
$$C = -1$$
$$y = \frac{-1}{\frac{x^2}{2} - 1} = \frac{-2}{x^2 - 2}$$

5d. The functions will have vertical asymptotes where the denominator has real roots. This will occur when $C \leq 0$.

6a. The function, G, is concave down when its derivative, f, is decreasing. This occurs on the interval $[-1, 1.5]$. On this interval, $G'' \leq 0$.

6b. From the graph $G'(0) = f(0) = 2$ is the slope of the tangent line. Therefore, $m = 2$ Since the y-intercept of the tangent line is the point $(0, 7)$, this must also be the point of tangency. Therefore $G(0) = 7$.

6c. Since the average value of f on $0 \leq x \leq 3$ is $\dfrac{1}{3-0} \int_0^3 f(x)\, dx = 0$, $\int_0^3 f(x)\, dx = 0$

From (b), $G(0) = 7$

$$G(3) = G(0) + \int_0^3 f(x)\, dx$$
$$= 7 + 0$$
$$= 7$$

Alternatively, one could reason that since the average value on $[0, 3]$ is zero the total change in the value between 0 and 3 is zero and therefore $G(3) = G(0) = 7$. This is borne out by the symmetry of the graph of f between 0 and 3.

Another approach is this:

$$G(3) = G(-2) + \int_{-2}^3 f(x)\, dx$$
$$= \underbrace{G(-2) + \int_{-2}^0 f(x)\, dx}_{= G(0)} + \underbrace{\int_0^3 f(x)\, dx}_{= 0}$$
$$= 7 + 0$$
$$= 7$$

Sample Examination II

1. $\int_0^2 (2x^3 + 3)\ dx = \left(\dfrac{2x^4}{4} + 3x\right)\Big|_0^2$

   $ = \left(\dfrac{x^4}{2} + 3x\right)\Big|_0^2$

   $ = (8+6) - (0+0) = 14$

   The correct choice is (C).

2. $\dfrac{5x-2}{(x-7)(x+4)} = \dfrac{A}{x-7} + \dfrac{B}{x+4} = \dfrac{A(x+4) + B(x-7)}{(x-7)(x+4)}$

   Then solve $5x - 2 = A(x+4) + B(x-7)$ by substituting $x = 7$;

   $5(7) - 2 = A(7+4) + B(7-7)$

   $33 = 11A \Rightarrow A = 3$

   Therefore, $\dfrac{3}{x-7}$ is one of the fractions obtained.

   The correct choice is (C).

3. $\dfrac{dx}{dt} = 2t$ and $\dfrac{dy}{dt} = -3t^2$

   $\vec{v_t} = (2t, -3t^2)$ and $\vec{a_t} = (2, -6t)$

   At $t = 1$, $\vec{a_t} = (2, -6)$.

   The correct choice is (B).

4. $$f(x) = (2+3x)^4$$
   $$f'(x) = 4(2+3x)^3(3)$$
   $$f''(x) = 3 \cdot 4(2+3x)^2(3)^2$$
   $$f'''(x) = 2 \cdot 3 \cdot 4(2+3x)(3)^3$$
   $$f^{(4)}(x) = 1 \cdot 2 \cdot 3 \cdot 4 \cdot (3)^4 = 4!3^4$$

   If $y = (a+bx)^n$, then the nth derivative, $\dfrac{d^n y}{dx^n} = n!b^n$

   The correct choice is (C).

5. $f(x) = x^4 - 8x^2$

   $f'(x) = 4x^3 - 16x = 4x(x^2 - 4)$

   $f'(x) = 0$ when $x = 0, -2,$ and 2.

	$x < -2$	$-2 < x < 0$	$0 < x < 2$	$x > 2$
f'	$-$	$+$	$-$	$+$

   $f(x)$ has a relative minimum where $f'(x)$ switches sign from $-$ to $+$.

   Therefore, $f(x)$ has a relative minimum at $x = 2$ and $x = -2$ only.

   The correct choice is (D).

6. $\displaystyle\int \sqrt{x}(x+2)\, dx = \int \left(x^{\frac{3}{2}} + 2x^{\frac{1}{2}}\right) dx$

   $\displaystyle = \frac{2}{5}x^{\frac{5}{2}} + 2 \cdot \frac{2}{3}x^{\frac{3}{2}} + C$

   $\displaystyle = \frac{2}{5}x^{\frac{5}{2}} + \frac{4}{3}x^{\frac{3}{2}} + C$

   The correct choice is (A).

7. $\dfrac{dy}{dx} = \dfrac{dy/dt}{dx/dt} = \dfrac{2t-1}{2t+1}$

   The tangent line to the curve is horizontal when $\dfrac{dy}{dx} = 0$ or when $\dfrac{2t-1}{2t+1} = 0 \Rightarrow t = \dfrac{1}{2}$.

   The correct choice is (D).

8. If the function $y = x^4 + bx^2 + 8x + 1$ has a horizontal tangent for some value of x, then the slope of y, y', is equal to 0 for that value of x.

   $y' = 4x^3 + 2bx + 8 \Rightarrow 4x^3 + 2bx + 8 = 0$

   If the function y has a point of inflection for some value of x, then $y'' = 0$ for that value of x.

   $y'' = 12x^2 + 2b \Rightarrow 12x^2 + 2b = 0 \Rightarrow 2b = -12x^2$

   Substituting $-12x^2$ for $2b$ in y',

   $$4x^3 - 12x^3 + 8 = 0$$
   $$-8x^3 = -8$$
   $$x^3 = 1$$
   $$x = 1$$

   Since $x = 1$, $2b = -12x^2$ or $2b = -12$, $b = -6$.

   Therefore the value of b is -6.

   The correct choice is (A).

9. Euler's method iterates the formulas

$$x_{n+1} = x_n + \Delta x \text{ and } y_{n+1} = y_n + \frac{dy}{dx}\Delta x$$

$$y(5.5) \approx 1 + (1-5)(0.5) = -1$$

$$y(6) \approx -1 + (-1-5.5)(0.5) = -4.25$$

The correct choice is (A).

10. The interval of convergence certainly includes 0.5 but not 1.5, so I is true and II is false. All Taylor polynomials are equal to the function at the center of the interval of convergence, so III is true.

The correct choice is (D).

11. Substitutions give the indeterminate form $\frac{0}{0}$, so using L'Hôpital's Rule:

$$\lim_{x\to 2} \frac{x^2-4}{\int_2^x \cos(\pi t)\, dt} = \lim_{x\to 2} \frac{2x}{\cos(\pi x)} = \frac{2(2)}{\cos(2\pi)} = \frac{4}{1} = 4$$

Note that by the Fundamental Theorem of Calculus: $\frac{d}{dx}\int_2^x \cos(\pi t)\, dt = \cos(\pi x)$.

The correct choice is (D).

12.   I. $\displaystyle\int_0^\infty e^{-x}\, dx = \lim_{t\to+\infty}\int_0^t e^{-x}\, dx = \lim_{t\to+\infty}(-e^{-x})\Big|_0^t = \lim_{t\to+\infty}(1-e^{-t}) = 1$

    II. $\displaystyle\int_0^1 \frac{1}{x^2}\, dx = \lim_{t\to 0^+}\int_t^1 \frac{1}{x^2}\, dx = \lim_{t\to 0^+}\left(-\frac{1}{x}\right)\Big|_t^1 = \lim_{t\to 0^+}\left(-1+\frac{1}{t}\right) \to +\infty;$

         The integral diverges.

    III. $\displaystyle\int_0^1 \frac{1}{\sqrt{x}}\, dx = \lim_{t\to 0^+}\int_t^1 \frac{1}{\sqrt{x}}\, dx = \lim_{t\to 0^+} 2\sqrt{x}\,\Big|_t^1 = \lim_{t\to 0^+} 2(1-\sqrt{t}) = 2$

Therefore, only I and III converge.

The correct choice is (E).

13. $x + y = xy$

By implicit differentiation, $1 + \dfrac{dy}{dx} = x\left(\dfrac{dy}{dx}\right) + y$

Solving for $\dfrac{dy}{dx}$, $\quad 1 - y = x\left(\dfrac{dy}{dx}\right) - \dfrac{dy}{dx}$

$$1 - y = \dfrac{dy}{dx}(x - 1)$$

$$\dfrac{dy}{dx} = \dfrac{1 - y}{x - 1}$$

The correct choice is (B).

14. The complete derivative with the correct Chain Rule inclusion is only found in choice (E). Let $u = g(x)$ and $du = g'(x)\,dx$.

The correct choice is (E).

15. Integrate the velocity function $v(t)$ to find the position function $y(t)$:

$$y(t) = \int v(t)\,dt = \int (8 - 2t)\,dt = 8t - t^2 + C$$

The velocity is positive when $t < 4$. During this time the particle is moving upwards.

At $t = 4$, the particle stops moving up and begins moving down, therefore $y(4) = 0$.

$$y(4) = 8(4) - 4^2 + C = 0 \Rightarrow C = -16$$

Therefore, the position function is $y(t) = -t^2 + 8t - 16$.

The correct choice is (A).

16. Using the substitution $u = \sqrt{x - 1}$,

$$u^2 = x - 1 \Rightarrow x = u^2 + 1$$

$$2u\,du = dx$$

Using $u = \sqrt{x - 1}$ to change the limits of integration,

when $x = 2$, $u = 1$, and when $x = 5$, $u = 2$.

So $\displaystyle\int_2^5 \dfrac{\sqrt{x-1}}{x}\,dx = \int_1^2 \dfrac{u}{u^2 + 1} \cdot 2u\,du = \int_1^2 \dfrac{2u^2}{u^2 + 1}\,du$

The correct choice is (E).

17. $e^x = 1 + x + \dfrac{x^2}{2!} + \cdots$

$e^{-x} = 1 - x + \dfrac{x^2}{2!} - \cdots$

$xe^{-x} = x\left(1 - x + \dfrac{x^2}{2!} - \cdots\right) = x - x^2 + \dfrac{x^3}{2!} - \cdots$

The correct choice is (B).

18. $\displaystyle\int_0^2 \left(2x^3 - kx^2 + 2k\right)\,dx = \dfrac{2x^4}{4} - \dfrac{kx^3}{3} + 2kx\bigg|_0^2 = \left(\dfrac{32}{4} - \dfrac{8k}{3} + 4k\right) - \left(0\right) = 8 - \dfrac{8k}{3} + 4k$

Since $\displaystyle\int_0^2 (2x^3 - kx^2 + 2k)\,dx = 12$,

therefore, $8 - \dfrac{8k}{3} + 4k = 12$

$24 + 4k = 36$

$k = 3$

The correct choice is (E).

19. $\displaystyle\sum_{n=1}^{\infty} \left(\dfrac{1}{2}\right)^{2n}$ is a geometric series, i.e. $\left(\dfrac{1}{2}\right)^2 + \left(\dfrac{1}{2}\right)^4 + \left(\dfrac{1}{2}\right)^6 + \cdots$

where the first term is $\left(\dfrac{1}{2}\right)^2$ or $\dfrac{1}{4}$ and the ratio is also $\left(\dfrac{1}{2}\right)^2$ or $\dfrac{1}{4}$.

The sum of a geometric series is $\dfrac{a}{1-r}$ where a is the first term in the series, and r is the ratio.

Therefore, $\displaystyle\sum_{n=1}^{\infty} \left(\dfrac{1}{2}\right)^{2n}$ is the sum of the geometric series, which is $\dfrac{\frac{1}{4}}{1 - \frac{1}{4}} = \dfrac{1}{3}$

The correct choice is (A).

20. In general, the derivative of $y = \ln u$ is $y' = \dfrac{du}{u}$.

The derivative of $y = \ln\sqrt{1-x^2}$ is

$y' = \dfrac{1}{\sqrt{1-x^2}} \cdot d\left(\sqrt{1-x^2}\right) = \dfrac{1}{\sqrt{1-x^2}} \cdot \dfrac{1}{2} \cdot \dfrac{1}{\sqrt{1-x^2}} \cdot -2x = \dfrac{-x}{1-x^2}$

The correct choice is (B).

26 Sample Examination II

21. "A function whose derivative is a constant multiple of itself" means

$$\frac{dy}{dx} = ky$$
$$\frac{dy}{y} = k\, dx$$
$$\ln|y| = k\,x + C_1$$
$$|y| = e^{kx+C_1} = e^{C_1}e^{kx}$$
$$y = C\,e^{kx}$$

This is an exponential function.

The correct choice is (E).

22. The graph is concave downward when $y'' < 0$.

$y = x^3 - 6x^2$, $y' = 3x^2 - 12x$, $y'' = 6x - 12$. $y'' < 0$ when $6x - 12 < 0$ or $x < 2$.

Hence, $y = x^3 - 6x^2$ is concave downward when $x < 2$.

The correct choice is (B).

23. Let $y =$ amount of substance present at time t.

You are given that $y(-4) = 12$ and $y(0) = 8$. You want to find $y(8)$.

$y' = ky \Rightarrow y = Ae^{kt} \Rightarrow y = 8e^{kt}$ (since $y(0) = 8$).

Since $y(-4) = 12$, $y = 8e^{kt} \Rightarrow 12 = 8e^{-4k} \Rightarrow \frac{3}{2} = e^{-4k} \Rightarrow 3e^{4k} = 2 \Rightarrow e^{4k} = \frac{2}{3} \Rightarrow$

$4k = \ln\frac{2}{3} \Rightarrow k = \frac{1}{4}\ln\frac{2}{3}$.

Therefore, $y = 8e^{(\frac{1}{4}\ln\frac{2}{3})t}$, so $y(8) = 8e^{\ln(\frac{2}{3})^2} = \frac{32}{9}$.

Alternative method: After 4 years $\frac{2}{3}$ of the amount remains. So after 4 more years $\frac{2}{3} \cdot 8 = \frac{16}{3}$ g. remain. After another 4 years $\frac{2}{3} \cdot \frac{16}{3} = \frac{32}{9}$ g. remain.

The correct choice is (C).

24. I. The series may be written as $1 + \frac{1}{2^{3/2}} + \frac{1}{3^{3/2}} + \cdots + \frac{1}{n^{3/2}} + \cdots$

This series is convergent because it is a p-series with $p > 1$.

II. Compare the given series with the convergent geometric series ($r = 1/2$),

$$\frac{1}{2} + \frac{1}{4} + \frac{1}{8} + \cdots + \frac{1}{2n} + \cdots.$$

Each term of the given series is equal to or less than the corresponding term of the convergent geometric series. Therefore the given series is convergent.

(The p-series, $\frac{1}{n^2}$ could also be used in the Comparison Test.)

III. Compare the given series with $1 + \frac{1}{2} + \frac{1}{3} + \cdots + \frac{1}{n} + \cdots$ Each term of the given series is equal to or greater than the corresponding term of the divergent harmonic series, therefore this series also diverges.

The correct choice is (C).

25. $f'(x) = \sum_{n=0}^{\infty} \frac{(-1)^{n+1}(2n+1)x^{2n}}{(2n+1)!} = \sum_{n=0}^{\infty} \frac{(-1)^{n+1}x^{2n}}{(2n)!}$, since $(2n+1)! = (2n+1)(2n)!$

The correct choice is (B).

26. $y = \sqrt[3]{x^2 - 1} = (x^2 - 1)^{\frac{1}{3}}$

$y' = \frac{1}{3}(x^2 - 1)^{-\frac{2}{3}}(2x)$

$y'(3) = (\frac{1}{3})(\frac{1}{4})(6) = \frac{1}{2}$

The normal line to the curve will have a slope of -2.

The equation of the normal line to the curve $y = \sqrt[3]{x^2 - 1}$ at $(3, 2)$ is

$y - 2 = -2(x - 3)$ or $y + 2x = 8$

The correct choice is (D).

27. Since the velocity, $s'(t)$, and the acceleration, $s''(t)$, are both positive, the graph of $s(t)$ will be increasing $(s'(t) > 0)$ and concave upwards $(s''(t) > 0)$. The graph of (C) represents $s(t)$.
The correct choice is (C).

28. If $f(x) = x^2$ on $[0, 1]$ and $\Delta x = \frac{1-0}{n} = \frac{1}{n}$,

then $S_n = \frac{1}{n}\left[\left(\frac{1}{n}\right)^2 + \left(\frac{2}{n}\right)^2 + \ldots + \left(\frac{n-1}{n}\right)^2\right]$ is a right-hand Riemann sum, and

$\lim\limits_{n \to \infty} S_n = \int_0^1 x^2 \, dx$.

The correct choice is (B).

29. Differentiate the position equations to find the velocity vector:

$\vec{v}(t) = (-9\sin t, 4\cos t)$.

Then differentiate the velocity vector to find the acceleration vector:

$\vec{a}(t) = (-9\cos t, -4\sin t)$

Substitute the value $t = 3$, $\vec{a}(3) = (-9\cos(3), -4\sin(3)) = (8.910, -0.564)$.

The correct choice is (C).

30. The graph of $f'(x)$ is increasing at $x = 2$, so its derivative $f''(2) > 0$

From the graph $f'(2) = 0$.

Since the derivative changes from negative to positive at $x = 2$, $f(2)$ is a minimum value. So $f(2) < f(1)$, and therefore, $f(2) < 0$.

So arranging the numbers in order $f(2) < f'(2) < f''(2)$.

The correct choice is (A).

31. Since $f'(x)$ changes sign from negative to positive at $x = 0$, the First Derivative Test indicates that $x = 0$ is a relative minimum. (A) must be true.

Since $f'(x)$ is always increasing, $f(x)$ is always concave upwards and has no point of inflection. Hence, (B) and (C) are not true.

Since $f(0)$ is not given you cannot determine if $f(x)$ passes through the origin — therefore, (D) need not be true.

Since the derivative $f'(x)$ is not even $\big((f'(10) \neq f'(-10)\big)$, $f(x)$ cannot be odd. The derivative of an odd function must be even. So (E) is not true.

The correct choice is (A).

32. $C = x(15 - x) - x = 14x - x^2$

    To maximize the catch, C, find C' and set it equal to zero:

    $C' = 14 - 2x$

    $C' = 0$ when $x = 7$

    When $x < 7$, $C' > 0$; when $x > 7$, $C' < 0 \Rightarrow x = 7$ is a maximum.

    Or, by the Second-Derivative Test, $C''(7) = -2$, indicating $x = 7$ is a maximum.

    The correct choice is (B).

33. To the right of the origin the function is decreasing; therefore, its derivative $f'(x) < 0$, and I is true. As $x \to +\infty$ and $x \to -\infty$, the graph flattens out so its slope $f'(x)$ is approaching zero; II is true and III is false.

    The correct choice is (D).

34. Integrate the velocity vector to obtain the position vector: $(2t^2 + C, -t^2 + K)$. Then graph this on the graphing calculator. Regardless of the constants of integration (try any values you want) this is a ray.

    Alternatively, $\dfrac{dy}{dx} = \dfrac{\frac{dy}{dt}}{\frac{dx}{dt}} = \dfrac{-2t}{4t} = -\dfrac{1}{2}$ is constant so the curve must be either a line or a ray. Since t is limited to non-negative numbers, the curve is a ray.

    The correct choice is (E).

35. The acceleration becomes negative when the velocity stops increasing and starts decreasing (even though the velocity may still be positive). This corresponds to a point of inflection on the graph, point C, where the concavity changes from up (on the left) to down (on the right).

    The correct choice is (C).

36. Graph $f'(x)$ in a suitable window such as $[0, 5]$ by $[-10, 10]$. The derivative changes from negative to positive at $x = 0.618$ indicating a relative minimum there. After the maximum at $x = 1.623$, the function decreases again making the endpoint at $x = 5$ a candidate for the absolute minimum also. Since the "negative area" below the axis after $x = 1.623$ is obviously so much larger than the small positive area, the function has lost much more than it gained since the relative minimum. Thus, the absolute minimum occurs when $x = 5$.

The correct choice is (E).

37. The derivative may be approximated by calculating the slope from any two points near $x = 4$, for example,
$$f'(4) \approx \frac{1.16016 - 1.15782}{4.00100 - 4.00000} = \frac{0.00234}{0.001} = 2.34$$
Using other points gives similar results.

The correct choice is (D).

38. A rational function may have only one horizontal asymptote. If it does, then as $x \to \pm\infty$ the function must approach this value. This is expressed by the limit in (B). For example, the graph of $f(x) = \frac{7x^2}{x^2+3}$ is such a function. Its graph and asymptote are shown below:

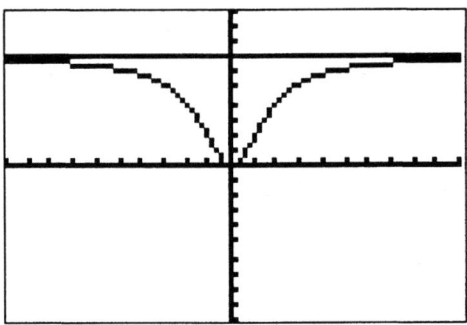

The correct choice is (B).

39. Graph the two functions $y = \cos 2x$ and $y = \sin 3x$ in the interval $0 \leq x \leq 5$. Choices (A), (B), (C) and (E) represent the areas of the regions enclosed by the graphs from left to right. It is easy to see that choice (E) is the largest. Choice (D) has the wrong function "on top" and therefore represents a negative number.

The correct choice is (E).

40. First graph the function $f(x)$ in the window $[0, 1.4]$ by $[-10, 10]$. Identify the zero at $x = 1.042$. Then use the derivative function to find the derivative value. The derivative at $x = 1.042$ is 3.451. Do not be misled by the vertical asymptote near $x = 0.411$.

The correct choice is (C).

41. The interval $[0, 6]$ is divided into three subintervals $[0, 2]$, $[2, 4]$, and $[4, 6]$. The integral is approximated using the function's value at the midpoint of each subinterval multiplied by the width of the subinterval:
$$(0.25)(2) + (0.68)(2) + (0.95)(2) = 3.76$$
For additional practice: (A) is the left-hand Riemann sum, (E) is the right-hand Riemann sum and (B) is the Trapezoidal Rule, all with $n = 3$.

The correct choice is (D).

42. For any number a,

$$\frac{d}{dx} \int_x^{x^3} \sin(t^2)\, dt = \frac{d}{dx} \left[\int_x^a \sin(t^2)\, dt + \int_a^{x^3} \sin(t^2)\, dt \right]$$

$$= \frac{d}{dx} \left[\int_a^{x^3} \sin(t^2)\, dt - \int_a^x \sin(t^2)\, dt \right]$$

$$= 3x^2 \sin\left((x^3)^2\right) - \sin(x^2)$$

$$= 3x^2 \sin(x^6) - \sin(x^2)$$

The correct choice is (C).

43. The amount of water, A, after x hours, is given by the accumulation function

$$A(x) = \int_0^x 300\sqrt{t}\, dt$$

After 4 hours, there are $\int_0^4 300\sqrt{t}\, dt$ gallons.

$$\int_0^4 300\sqrt{t}\, dt = 300 \int_0^4 t^{\frac{1}{2}}\, dt = 300 \cdot \frac{2}{3} \cdot t^{\frac{3}{2}} \Big|_0^4 = 200\left(4^{\frac{3}{2}} - 0\right) = 1600 \text{ gallons}$$

The correct choice is (D).

44. Solve $2\sin x = x$ to obtain $x = 1.895$.

Then evaluate $V_x = \pi \int_0^{1.895} \left[(2\sin x)^2 - x^2 \right] dx = 6.678$ with a graphing calculator.

The correct choice is (D).

45.

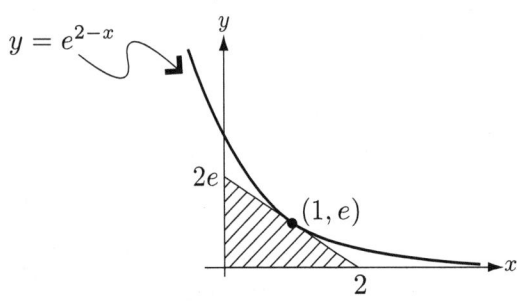

First, find the equation of the tangent line to $y = e^{2-x}$ at $(1, e)$. $y' = -e^{2-x}$; therefore, at $x = 1$, $y'(1) = -e$. The equation of the tangent line is $y - e = -e(x - 1)$ or $y = -ex + 2e$. To find the area of the triangle (the shaded region) you need the base and height of the triangle; these are the x-intercept and the y-intercept of the tangent line. The x-intercept is $x = 2$ and the y-intercept is $y = 2e$.

Therefore, the area of the triangle $= \frac{1}{2}$ (base) (height) $= \frac{1}{2}(2)(2e) = 2e$.

The correct choice is (A).

Sample Examination II

1a. The distance is given by $\int_0^5 (-5t^2 + 20t + 25)\, dt = 166.666$ miles.

1b. The acceleration is the derivative of the velocity:

$$\left.\frac{d}{dt}v(t)\right|_{t=3} = -10t + 20\Big|_{t=3} = -10(3) + 20 = -10 \text{ miles/hour/hour}$$

1c. Any one of the three approaches below is reasonable. Be sure to show your work so that the method you use is clear.

Using a <u>left-hand sum</u> with $n = 5$, the distance traveled in the second part of the trip is (1 hour)$(0 + (-25) + (-20) + (-50) + A$ miles / hour) $= -95 + A$ miles. So $-95 + A \approx -167$ and $A \approx -72$ miles per hour.

Using a <u>right-hand sum</u> with $n = 5$, gives (1 hour)$((-25) + (-20) + (-50) + A + (-20)$ miles per hour) $= -115 + A$ miles. So $-115 + A \approx -167$ and $A \approx -52$ miles per hour.

The <u>Trapezoidal Rule</u>, $T = \dfrac{L+R}{2}$, gives $\dfrac{(-95+A) + (-115+A)}{2} \approx -167$ and $A \approx -62$ miles per hour.

Sample Examination II

2a. Graph the function $T(x)$ in a suitable window such as $[0, 24]$ by $[50, 100]$. Use your calculator to find the value of x which makes $y = 70$. Rounded to the nearest half-hour, $x \approx 8.5$ or at 8:30 am.

2b. Same idea as in part (a). Use your calculator again to find the value of x which makes $y = 77$. Rounded to the nearest half-hour, $x \approx 10.5$ or at 10:30 am.

2c. Use the cost equation with $x = 8.5$ and $T_0 = 70$:

$$C(x) = 0.16 \int_{8.5}^{18} (T(x) - 70) \, dx = 0.16 \int_{8.5}^{18} \left[73 - 14 \cos\left(\frac{\pi(x - 3.4)}{12}\right) - 70 \right] dx = \$18.26$$

Of course, use your graphing calculator to evaluate the definite integral.

2d. In a like manner, use $T_0 = 77$ and $x = 10.5$ to first find the cost of cooling at $77°F$:

$$C(x) = 0.16 \int_{10.5}^{18} \left[73 - 14 \cos\left(\frac{\pi(x - 3.4)}{12}\right) - 77 \right] dx = \$8.79$$

The savings is $\$18.26 - \$8.79 = \$9.47$ per day.

3a. Graph

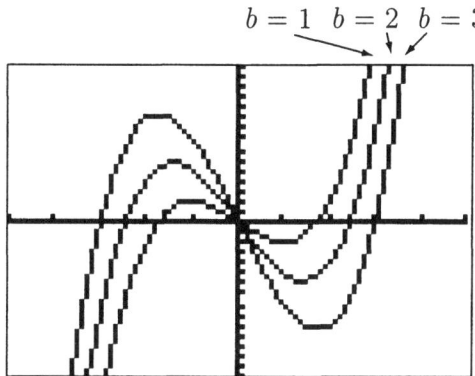

3b. The critical points occur where the derivative is zero:

$$f'(x) = 3x^2 - 3b$$

$$f'(x) = 0 \text{ when } x = \pm\sqrt{b}$$

From the First Derivative Test, the relative maximum points occur to the left of the origin or when $x = -\sqrt{b}$.

$$f(-\sqrt{b}) = (-\sqrt{b})^3 - 3b(-\sqrt{b}) = 2(\sqrt{b})^3 = 2b^{\frac{3}{2}}$$

The coordinates of the relative maximum points are $(-\sqrt{b}, 2b^{\frac{3}{2}})$.

3c. The relative minimum points occur to the right of the origin or when $x = +\sqrt{b}$.

$$f(+\sqrt{b}) = (\sqrt{b})^3 - 3b(\sqrt{b}) = -2(\sqrt{b})^3 = -2b^{\frac{3}{2}}$$

The coordinates of the relative minimum points are $(\sqrt{b}, -2b^{\frac{3}{2}})$

3d. Since the extrema (relative maximum and relative minimum points) lie on the graph of $y = -ax^3$, substitute their coordinates into this equation:

$$\pm 2b^{\frac{3}{2}} = -a(\pm\sqrt{b})^3 \text{ and solve for } a:$$

$$\frac{\pm 2b^{\frac{3}{2}}}{(\pm\sqrt{b})^3} = +2 = a$$

Therefore, $a = 2$; the relative maximum and relative minimum points will lie on the function $y = -2x^3$.

4a. $y = \dfrac{x}{x+C}, x \neq -C$

$y' = \dfrac{(x+C)(1) - x(1)}{(x+C)^2}$

$y' = \dfrac{C}{(x+C)^2}$

and substituting into the differential equation: $y' = \dfrac{y - y^2}{x}$

$$= \dfrac{\frac{x}{x+C} - \frac{x^2}{(x+C)^2}}{x}$$

$$= \dfrac{x(x+C) - x^2}{x(x+C)^2}$$

$$= \dfrac{C}{(x+C)^2}$$

4b. By substituting $(0,0)$ into $y = \dfrac{x}{x+C}$: $0 = \dfrac{0}{0+C}$ is true, so regardless of the value of C, $(0,0)$ is on the solution curve.

4c. Substituting $(-1, -1)$ into $y = \dfrac{x}{x+C}$ gives $-1 = \dfrac{-1}{-1+C}$ and $C = 2$. The particular solution is $y = \dfrac{x}{x+2}$, and the slope at $(0,0)$ is $y'(0) = \dfrac{2}{(0+2)^2} = \dfrac{1}{2}$.

Note that at $(0,0)$ the derivative in the form $\dfrac{dy}{dx} = \dfrac{y - y^2}{x}$ is an indeterminate form. The other form of the derivative must be used to find the value.

4d. Note that there are asymptotes of $x = -2$ and $y = 1$.

5a. $\vec{v} = (x'(t), y'(t)) = (2, -2t)$

5b. $\dfrac{dy}{dx} = \dfrac{dy/dt}{dx/dt} = \dfrac{-2t}{2} = -t$ is the slope at any point. Then the equation is

$y - (36 - t^2) = (-t)(x - 2t)$ or $y = -t(x - 2t) + (36 - t^2)$

5c. The line from part (b) intersects the x-axis when $y = 0$. Substitute this value into the equation of the tangent line and solve for x:

$$0 = -tx + 2t^2 + 36 - t^2$$
$$tx = t^2 + 36$$
$$x = \dfrac{t^2 + 36}{t} = t + 36t^{-1}$$

5d. The speed will be the slowest when $\left|\dfrac{dx}{dt}\right|$, obtained from part (c), is the smallest.

$\left|\dfrac{dx}{dt}\right| = \left|1 - \dfrac{36}{t^2}\right| = \dfrac{36 - t^2}{t^2}$. Therefore when $t = 6$ the light will be moving along the x-axis at its slowest speed (zero).

6a. At any point the slope is $\frac{df}{dx} = -2x$. The tangent line's equation is

$$y - (4 - a^2) = -2a(x - a)$$
$$y = -2ax + 2a^2 + 4 - a^2$$
$$y = -2ax + a^2 + 4$$

6b. Using the answer from (a) the area is the integral of the upper curve (the line) minus the lower:

$$A = \int_0^a \left((-2ax + a^2 + 4) - (4 - x^2)\right) dx$$

6c. Evaluating the integral in (b):

$$A = \int_0^a \left((-2ax + a^2 + 4) - (4 - x^2)\right) dx$$
$$A = -ax^2 + a^2 x + 4x - 4x + \frac{1}{3}x^3 \Big|_0^a$$
$$A = -a^3 + a^3 + 4a - 4a + \frac{1}{3}a^3$$
$$A = \frac{1}{3}a^3$$

6d. $A(x) = \frac{1}{3}x^3$

The average value of the area on the interval $0 \leq x \leq a$ is

$$\frac{1}{a - 0}\int_0^a \frac{1}{3}x^3 \, dx = \frac{1}{12a}x^4 \Big|_0^a = \frac{1}{12a}\left(a^4 - 0\right) = \frac{a^3}{12}$$

Sample Examination III

1. The area of the region is represented by $\int_1^3 (3x^2 + 2x)\, dx$.

   $\int_1^3 (3x^2 + 2x)\, dx = x^3 + x^2 \Big|_1^3 = (27 + 9) - (1 + 1) = 36 - 2 = 34$

   The correct choice is (B).

2. Since $\cos 2t = 2\cos^2 t - 1 \Rightarrow y = 1 - \cos 2t$
   $$y = 1 - (2\cos^2 t - 1)$$
   $$y = 2 - 2\cos^2 t$$
   $$y = 2 - 2x^2 \quad (\text{since } x = \cos t)$$

   which is part of a parabola (which traces and retraces the parabolic arch).

   The correct choice is (B).

3. To find the point of inflection, set $y'' = 0$ and check the concavity at that point, making sure that it switches from concave up to concave down or vice versa.

   $y' = (2x)(-e^{-x}) + 2e^{-x} = -2xe^{-x} + 2e^{-x} = -2e^{-x}(x - 1)$

   $y'' = (-2e^{-x}) + (x - 1)(2e^{-x}) = 2xe^{-x} - 4e^{-x} = 2e^{-x}(x - 2)$

   $y'' = 0$ only when $x = 2$, since e^{-x} is always positive.

   When $x < 2$, $y'' < 0$ and when $x > 2$, $y'' > 0$. This indicates that the only point of inflection is at $x = 2$.

   The correct choice is (D).

4. The velocity vector is found by finding $\dfrac{dx}{dt}$ and $\dfrac{dy}{dt}$.

   $\dfrac{dx}{dt} = (e^t)(\cos t) + (\sin t)(e^t)$ and $\dfrac{dy}{dt} = (e^t)(-\sin t) + (\cos t)(e^t)$

   At $t = \pi$, $\dfrac{dx}{dt} = (e^\pi)(\cos \pi) + (\sin \pi)(e^\pi) = -e^\pi + 0 = -e^\pi$

At $t = \pi$, $\dfrac{dy}{dt} = (e^\pi)(0) + (-1)(e^\pi) = -e^\pi$

Therefore, the velocity vector $\vec{v}_t = (-e^\pi, -e^\pi)$.

The correct choice is (E).

5. The velocity is positive so the particle does not change direction; the distance a particle travels from $t = 1$ to $t = 3$ can be represented by

$$\int_1^3 v(t)\, dt = \int_1^3 t^2\, dt = \dfrac{t^3}{3}\bigg|_1^3 = 9 - \dfrac{1}{3} = \dfrac{26}{3}$$

The correct choice is (C).

6. Using the Product Rule the derivative is

$$4x^2(-\sin(x)) + (\cos(x))(8x) = 8x\cos(x) - 4x^2\sin(x)$$

The correct choice is (B).

7. The choices give three different initial conditions for the unknown differential equation. For each trace along the slope field and see where the curves go. Starting above $(0, 2)$ the path moves down to the right leveling off near $y = 2$. So I is true. Starting between 0 and 2 on the y-axis and moving to the right the curve rises towards $y = 2$, so II is true. Starting on the y-axis below the origin and tracing to the right leads away from the axis, so III is false.

The correct choice is (D).

8. Let $u = x^2 + 1$

$du = 2x\, dx \Rightarrow x\, dx = \dfrac{1}{2} du$

$$\int \dfrac{x}{x^2 + 1}\, dx = \dfrac{1}{2}\int \dfrac{du}{u} = \dfrac{1}{2}\ln|u| = \dfrac{1}{2}\ln(x^2 + 1)\bigg|_1^3 = \dfrac{1}{2}(\ln 10 - \ln 2)$$

Since $\ln 10 - \ln 2 = \ln\left(\dfrac{10}{2}\right) = \ln 5$, therefore, $\int_1^3 \dfrac{x}{x^2+1}\, dx = \dfrac{1}{2}\ln 5$.

The correct choice is (D).

9. Using the properties of logarithms and the Chain Rule:

$$\frac{d}{dx} \ln\left(\frac{1}{x^2-1}\right) = \frac{d}{dx}\left(\ln 1 - \ln(x^2-1)\right)$$
$$= \frac{d}{dx}\left(-\ln(x^2-1)\right)$$
$$= -\frac{2x}{x^2-1}$$

The correct choice is (B).

10. $\int_1^\infty x^{-\frac{5}{4}}\,dx$ is an improper integral.

$$\int_1^\infty x^{-\frac{5}{4}}\,dx = \lim_{t\to\infty}\int_1^t x^{-\frac{5}{4}}\,dx = \lim_{t\to\infty}\left(\frac{x^{-\frac{1}{4}}}{-1/4}\right)\bigg|_1^t$$
$$= \lim_{t\to\infty}\left(\frac{t^{-\frac{1}{4}}}{-1/4} - \frac{1^{-\frac{1}{4}}}{-1/4}\right)$$
$$= 0 - \frac{1}{-1/4} = 4$$

The correct choice is (A).

11. $u = x^2 + 1 \Rightarrow x^2 = u - 1$

$du = 2x\,dx \Rightarrow x\,dx = \frac{1}{2}\,du$

When $x = 0, u = 1; x = 2, u = 5$

$$\int_0^2 \frac{x^3}{x^2+1}\,dx = \int_0^2 \frac{x^2}{x^2+1}\cdot x\,dx = \int_1^5 \frac{(u-1)}{u}\cdot\frac{1}{2}\,du = \int_1^5 \frac{u-1}{2u}\,du$$

The correct choice is (A).

12. Method I:

$$y' = \frac{k(k+x) - (kx+8)}{(k+x)^2}, \text{ or}$$

$$y' = \frac{k^2 - 8}{(k+x)^2}$$

Since $y = x + 4$ is the line tangent to y at $x = -2$, its slope is $y' = 1$.

By substituting $x = -2$ and $y'(-2) = 1$,

$$1 = \frac{k^2 - 8}{(k-2)^2} \Rightarrow k^2 - 8 = (k-2)^2 \text{ or } k^2 - 8 = k^2 - 4k + 4,$$

and $k = 3$.

Method II:

From the tangent line at $x = -2$, $y = -2 + 4 = 2$.

Substitute the point of tangency into the function and

$$2 = \frac{-2k + 8}{k - 2}$$
$$2k - 4 = -2k + 8$$
$$4k = 12$$
$$k = 3$$

The correct choice is (D).

13. If $\int_0^k (4kx - 5k)\, dx = k^2$, then $\left. \frac{4kx^2}{2} - 5kx \right|_0^k = k^2$, or

$$\left. 2kx^2 - 5kx \right|_0^k = k^2$$
$$2k^3 - 5k^2 = k^2$$
$$2k^3 = 6k^2 \quad \text{(dividing both sides by } k^2\text{, since } k > 0\text{)}$$
$$2k = 6$$
$$k = 3$$

The correct choice is (C).

14. The series is a geometric series with a ratio of $\left(-\frac{\pi}{3}\right)$.

Since the absolute value of the ratio is larger than 1, the series will not converge.

The correct choice is (E).

15. In general, if $y = a^u$, then $y' = a^u \cdot \ln a \cdot du$

If $y = 5^{(x^3 - 2)}$, then $\frac{dy}{dx} = 5^{(x^3 - 2)} \cdot \ln 5 \cdot 3x^2$

The correct choice is (B).

16. The given expression is the right-hand Riemann sum for the function $\sin(\pi x)$ on the interval $[0, 1]$. Therefore its limit as $n \to \infty$ is

$$\int_0^1 \sin(\pi x)\,dx = \frac{1}{\pi}\int_0^1 \sin(\pi x)\pi\, dx = \left.-\frac{1}{\pi}\cos(\pi x)\right|_0^1 = \frac{1}{\pi}((-\cos(\pi) + \cos(0)) = \frac{1}{\pi}(-(-1) + 1) = \frac{2}{\pi}$$

The correct choice is (B).

17. You are given the graph of $f'(x)$ and you must obtain the graph of $f(x)$. First, observe that $f(x)$ is increasing where $f'(x) \geq 0$; $f(x)$ is decreasing where $f'(x) < 0$. In the given graph of $f'(x)$, $f'(x) \geq 0$ for $x \geq -2$; $f'(x) < 0$ for $x < -2$. Therefore, $f(x)$ is decreasing for $x < -2$ and increasing for all $x \geq -2$. From the given choices, the only graph that is decreasing to the left of $x = -2$ and increasing to the right of $x = -2$ is (D).

    The correct choice is (D).

18. $\dfrac{dy}{dt} = \dfrac{1}{2}(2t+5)^{-\frac{1}{2}} \cdot 2 = (2t+5)^{-\frac{1}{2}}$ and $\dfrac{dx}{dt} = 1 - 2t$

    $\dfrac{dy}{dx} = \dfrac{dy/dt}{dx/dt} = \dfrac{(2t+5)^{-\frac{1}{2}}}{1-2t}\bigg|_{t=2} = \dfrac{9^{-\frac{1}{2}}}{-3} = -\dfrac{1}{9}$

    The correct choice is (A).

19. The segment with endpoints $(-2, 26)$ and $(10, 2)$ has a slope of -2. Therefore, by the Mean Value Theorem, the line through $(4, 23)$ will be parallel to the segment and tangent to the graph; its equation is $y - 23 = -2(x - 4)$. Its y-intercept is 31.

    The correct choice is (C).

20. Since at each point (x, y) on a certain curve, the slope of the curve is $4xy$; $\dfrac{dy}{dx} = 4xy$.
    To find y (the equation of the curve), solve this separable differential equation.

    $$\dfrac{dy}{dx} = 4xy \Rightarrow \dfrac{dy}{y} = 4x\ dx \Rightarrow \int \dfrac{dy}{y} = \int 4x\ dx$$
    $$\Rightarrow \ln|y| = 2x^2 + C_1$$
    $$\Rightarrow y = Ce^{2x^2}$$

    Since the curve contains the point $(0, 4)$, $4 = Ce^0 \Rightarrow C = 4$.

    Therefore, $y = 4e^{2x^2}$.

    The correct choice is (E).

21. Using integration by partial fractions,

    $$\dfrac{1}{x^2-9} = \dfrac{1}{(x-3)(x+3)} = \dfrac{A}{x-3} + \dfrac{B}{x+3} \Rightarrow 1 = A(x+3) + B(x-3)$$

    Let $x = 3$. Then $1 = 6A$ or $A = \dfrac{1}{6}$.

Let $x = -3$. Then $1 = -6B$ or $B = -\frac{1}{6}$.

So $\dfrac{1}{x^2 - 9} = \dfrac{1}{6}\left(\dfrac{1}{x-3}\right) - \dfrac{1}{6}\left(\dfrac{1}{x+3}\right)$

Therefore, $\displaystyle\int \dfrac{dx}{x^2-9} = \dfrac{1}{6}\int \dfrac{dx}{x-3} - \dfrac{1}{6}\int \dfrac{dx}{x+3} = \dfrac{1}{6}\ln|x-3| - \dfrac{1}{6}\ln|x+3| + C = \dfrac{1}{6}\ln\left|\dfrac{x-3}{x+3}\right| + C$

The correct choice is (E).

22. The length of a continuous curve $y = f(x)$ from $x = a$ to $x = b$ is $L = \displaystyle\int_a^b \sqrt{1 + (y')^2}\, dx$.

In the given problem, $y = e^{(e^x)}$. So $y' = e^{(e^x)} \cdot e^x = e^{(x+e^x)}$ and $(y')^2 = e^{2(x+e^x)}$.

Therefore, $L = \displaystyle\int_0^1 \sqrt{1 + e^{2(x+e^x)}}\, dx$.

The correct choice is (A).

23. The rate of change of a number is the derivative of the number. Thus the rate of change of x is $\dfrac{dx}{dt} = x$ and the rate of change of the reciprocal of x is

$$\dfrac{d}{dt}\left(\dfrac{1}{x}\right) = -x^{-2}\dfrac{dx}{dt}$$
$$= -\left(\dfrac{1}{4}\right)^{-2}\left(\dfrac{1}{4}\right)$$
$$= -(16)\left(\dfrac{1}{4}\right)$$
$$= -4$$

The correct choice is (B).

24. $f(x) = \sqrt{e^{2x} + 1} = \left(e^{2x} + 1\right)^{\frac{1}{2}}$ and $f'(x) = \dfrac{1}{2}\left(e^{2x}+1\right)^{-\frac{1}{2}} \cdot 2e^{2x} = \dfrac{e^{2x}}{\sqrt{e^{2x}+1}}$

Therefore, $f'(0) = \dfrac{e^0}{\sqrt{e^0+1}} = \dfrac{1}{\sqrt{2}} = \dfrac{\sqrt{2}}{2}$

The correct choice is (C).

25. <u>Method 1</u>:

Notice that $x^2y + yx^2 = 2x^2y$.

So $2x^2y = 6 \Rightarrow y = \dfrac{6}{2x^2} = \dfrac{3}{x^2} = 3x^{-2}$

To find $\dfrac{d^2y}{dx^2}$, take the second derivative:

$y = 3x^{-2}$

$y' = -6x^{-3}$

$y'' = \dfrac{d^2y}{dx^2} = 18x^{-4}$ or $\dfrac{18}{x^4}$

So $\dfrac{d^2y}{dx^2}$ at $x = 1$ is $\dfrac{18}{1^4} = 18$

For this problem there was really no need for implicit differentiation.

<u>Method II</u>:

Using implicit differentiation, first find $\dfrac{dy}{dx}$ (first derivative):

$\left(x^2\dfrac{dy}{dx} + 2xy\right) + \left(2xy + x^2\dfrac{dy}{dx}\right) = 0 \Rightarrow 4xy + 2x^2\dfrac{dy}{dx} = 0 \Rightarrow \dfrac{dy}{dx} = \dfrac{-4xy}{2x^2} = \dfrac{-2y}{x}$

Now, since $\dfrac{dy}{dx} = \dfrac{-2y}{x}$, find $\dfrac{d^2y}{dx^2}$ by differentiating $\dfrac{-2y}{x}$ implicitly.

$\dfrac{d^2y}{dx^2} = \dfrac{x\left(-2\dfrac{dy}{dx}\right) - (-2y)}{x^2} = \dfrac{-2x\left(\dfrac{-2y}{x}\right) + 2y}{x^2} = \dfrac{6y}{x^2}$

<u>Note</u>: Substitute $\dfrac{-2y}{x}$ for $\dfrac{dy}{dx}$ in the equation above.

Therefore $\dfrac{d^2y}{dx^2} = \dfrac{6y}{x^2}$ and at $(1,3)$, $\dfrac{d^2y}{dx^2} = \dfrac{6(3)}{1^2} = 18$

The correct choice is (E).

26. Since every cross section is a semicircle, Area $= \frac{1}{2}\pi(\text{radius})^2 = \frac{1}{2}\pi\left(\frac{1}{2}y\right)^2 = \frac{\pi}{8}y^2$.

$$\text{Volume} = \int_0^4 \frac{\pi}{8}\left(\frac{4-x}{2}\right)^2 dx \text{ (since } x+2y=4 \Rightarrow 2y=4-x \text{ or } y=\frac{4-x}{2})$$

$$= \frac{\pi}{32}\int_0^4 (4-x)^2\, dx$$

$$= \frac{\pi}{32}\int_0^4 (16-8x+x^2)\, dx$$

$$= \frac{\pi}{32}\left(16x - 4x^2 + \frac{x^3}{3}\right)\Big|_0^4$$

$$= \frac{\pi}{32}\left[\left(64 - 64 + \frac{64}{3}\right) - (0)\right]$$

$$= \frac{\pi}{32}\left(\frac{64}{3}\right) = \frac{2\pi}{3}$$

The correct choice is (A).

27. I diverges because its partial sums alternate between 1 and 0.

II diverges because $\lim_{n\to\infty}(-1)^{n+1}n \neq 0$.

III diverges because $\lim_{n\to\infty}\left(\frac{1+n}{n}\right)^n = \lim_{n\to\infty}\left(1+\frac{1}{n}\right)^n = e$.

Since the limit of the nth term is not zero, the series diverges.

The correct choice is (A).

28. Simplify the expression and find the limit algebraically:

$$\lim_{x\to a}\frac{x^4-a^4}{x^2-a^2} = \lim_{x\to a}\frac{(x^2-a^2)(x^2+a^2)}{x^2-a^2}$$
$$= \lim_{x\to a}(x^2+a^2)$$
$$= 2a^2$$

Since the limit is given as 16 solve the equation:

$$2a^2 = 16$$
$$a^2 = 8$$
$$a = \sqrt{8}$$
$$a = 2\sqrt{2}$$

This limit may also be found using L'Hôpital's Rule.

The correct choice is (B).

29. In polar form r represents the distance from the origin (or pole). To determine how r is behaving consider $\frac{dr}{d\theta} = -5\sin\theta$.

$\frac{dr}{d\theta} < 0$ for $0 < \theta < \pi$ and $2\pi < \theta < 8$

$\frac{dr}{d\theta} > 0$ for $\pi < \theta < 8$.

Since the derivative changes from negative to positive at $\theta = \pi$ only, this is where the particle's distance from the origin starts to increase.

The correct choice is (B).

30. For values in the given interval, this is an alternating series. The error will be less than the first omitted term. The largest difference will occur at $x = \frac{\pi}{2}$, the value most distant from the center of the interval of convergence ($x = 0$). Solve

$$\left|(-1)^n \frac{\left(\frac{\pi}{2}\right)^n}{n!}\right| < 0.0001$$

on your calculator by evaluating the left side for the answer choices.

For the fifth term $n = 8$ and $\frac{\left(\frac{\pi}{2}\right)^8}{8!} \approx 0.00092$ is too large.

For the sixth term $n = 10$ and $\frac{\left(\frac{\pi}{2}\right)^{10}}{10!} \approx 0.000025 < 0.0001$.

So 6 terms are sufficient.

The correct choice is (C).

31. Find the second derivative of f:

$f'(x) = 2x - 5\sin x$

$f''(x) = 2 - 5\cos x$

Points of inflection occur where the second derivative changes sign. In this case, the second derivative is zero and changes sign when $\cos x = \frac{2}{5}$. In any interval of length 2π this happens twice. Therefore, in the given interval $0 \leq x \leq 2\pi n$, it will occur $2n$ times.

The correct choice is (E).

48   Sample Examination III

32. Since the function has a derivative for all $x \neq 2$, it is continuous everywhere else. The derivative is positive indicating that the function is always increasing. The function must increase to infinity on the left side of the asymptote and then increase $from -\infty$ on the right side. In other words, this is an odd vertical asymptote. Thus the limit in III should be $-\infty$ and the limit in I does not exist since the two one-sided limits are different. Only II is true. An example of such a function is $f(x) = \dfrac{-1}{x-2}$; its graph is shown below:

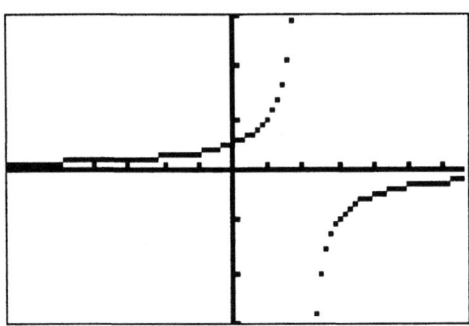

The correct choice is (B).

33. A graph of the derivative on the given interval reveals that the derivative has three zeros, indicating three critical points.

   The correct choice is (D).

34. A function $f(x)$ grows faster than $g(x)$ as $x \to \infty$ if $\lim\limits_{x \to \infty} \dfrac{f(x)}{g(x)} = \infty$. In this problem, a function, say $f(x)$, will grow faster than e^x as $x \to \infty$ if $\lim\limits_{x \to \infty} \dfrac{f(x)}{e^x} = \infty$.

   Examine the choices:

   (A) $\lim\limits_{x \to \infty} \dfrac{x^4}{e^x} = 0$ (Graph $\dfrac{x^4}{e^x}$ and see that as $x \to \infty$, $\dfrac{x^4}{e^x} \to 0$).

   Note: Exponential functions like 2^x and e^x grow faster than polynomial functions as $x \to \infty$. For that matter, e^x grows faster than any power of x, say $x^{1,000,000,000}$ as $x \to \infty$.

   (B) $\lim\limits_{x \to \infty} \dfrac{\ln x}{e^x} = 0$ (Graph $\dfrac{\ln x}{e^x}$ and see that as $x \to \infty$, $\dfrac{\ln x}{e^x} \to 0$).

   Note: $\ln x$ grows slower than any polynomial as $x \to \infty$, and (obviously) slower than e^x as $x \to \infty$

   (C) $\lim\limits_{x \to \infty} \dfrac{e^{-x}}{e^x} = \lim\limits_{x \to \infty} \left(\dfrac{1}{e^{2x}}\right) = 0$

(D) $\lim_{x\to\infty} \dfrac{3^x}{e^x} = \lim_{x\to\infty} \left(\dfrac{3}{e}\right)^x = \infty$. So 3^x grows faster than e^x as $x \to \infty$.

Keep in mind, that if $a > b > 0$, then a^x always grows faster than b^x as $x \to \infty$. Since $a > b > 0$, so $\left(\dfrac{a}{b}\right) > 1$, and $\lim_{x\to\infty} \dfrac{a^x}{b^x} = \lim_{x\to\infty} \left(\dfrac{a}{b}\right)^x = \infty$.

(E) $\lim_{x\to\infty} \dfrac{\frac{1}{2}e^x}{e^x} = \lim_{x\to\infty} \dfrac{1}{2} = \dfrac{1}{2}$.

f and g grow at the <u>same</u> rate as $x \to \infty$ if $\lim_{x\to\infty} \dfrac{f(x)}{g(x)} = L \neq 0$; that is, L is finite and not zero.

In this choice, the limit is $\dfrac{1}{2}$, so $f(x)$ grows at the same rate as $g(x)$, and $f(x)$ does not grow faster than $g(x)$ as $x \to \infty$.

Note: This problem could have been solved by graphing the functions in the window $[0, 10]$ by $[0, 6500]$. You will see that 3^x is the fastest growing function. Be careful: x^4 appears greater then e^x, but the functions intersect at $x \approx 8.6$. Remember, every exponential function a^x, with $a > 1$, grows faster than any polynomial function.

The correct choice is (D).

35. Graph the second derivative. The function $f(x)$ will have a point of inflection when the second derivative changes sign. The second derivative changes sign (crosses the x-axis) six times in this interval.

The correct choice is (C).

36.
$$A = \int_0^\pi \dfrac{1}{2}[2(\cos\theta + \sin\theta)]^2 \, d\theta$$
$$= \int_0^\pi 2(\cos^2\theta + \sin^2\theta + 2\sin\theta\cos\theta) \, d\theta$$
$$= \int_0^\pi 2(1 + \sin 2\theta) \, d\theta \quad (\text{since } 2\sin\theta\cos\theta = \sin 2\theta \text{ and } \sin^2\theta + \cos^2\theta = 1)$$
$$= \int_0^\pi (2 + 2\sin 2\theta) \, d\theta = 2\theta + (-\cos 2\theta)\Big|_0^\pi = 2\pi$$

The integration is done from 0 to π corresponding to once around the graph. The graph is traced twice as θ goes from 0 to 2π.

The correct choice is (D).

37. To approximate $f(2.2)$ center the Taylor polynomial about $x = 2$.

$$f(x) \approx f(2) + f'(2)(x-2) + \frac{f''(2)}{2!}(x-2)^2 + \frac{f'''(2)}{3!}(x-2)^3 \text{ for } x \text{ near 2}.$$

$$f(2.2) \approx 7.20 + 3.50(2.2-2) + \frac{1.75}{2}(2.2-2)^2 + \frac{0.85}{6}(2.2-2)^3$$

$$f(2.2) \approx 7.20 + 3.50(0.2) + \frac{1.75}{2}(0.2)^2 + \frac{0.85}{6}(0.2)^3$$

Keep in mind, that it is always best to center the polynomial closer to the x-value "asked for". That is, do not center this polynomial about $x = 0$.

The correct choice is (E).

38. Find $T(0.5) = 0.479167$ and $f(0.5) = \sin(0.5) = 0.479426$. The difference between these two numbers is $|E| = |0.479426 - 0.479167| = 0.000259$.

The correct choice is (B).

39. Begin by graphing this function in the window $[0, 1]$ by $[-2, 2]$. Then change XMAX to 0.1, then 0.01, then 0.001, etc. Each time you will see a different graph! As x decreases from 1 to zero, $\ln x$ decreases from zero to $-\infty$. The $\sin(\ln x)$ acts like the $\sin x$ as x goes from zero to $-\infty$, only it does it in the space from one to zero (like a compressed spring). The zeros occur when $\sin(\ln x) = 0$ which is when $\ln x = \pm k\pi$, $k = 0, 1, 2, 3, \ldots$, thus $x = e^{\pm k\pi}$, $k = 0, 1, 2, 3, \ldots$. All of the negative powers are in the given interval, therefore there are an infinite number of roots. HOWEVER, since each root is over an order of magnitude larger than the next smaller root, it is impossible to "see" more than three roots in any graphing window. You need to be aware of the limitations of the graphing calculator. In fact, this problem is probably best solved with a paper and pencil!

The correct choice is (E).

40. If $f^{-1}(x) = g(x)$, then $g'(x) = \dfrac{1}{f'(g(x))}$ and $g'(0) = \dfrac{1}{f'(g(0))}$.

Solve on your calculator $x^3 - 7x^2 + 25x - 39 = 0$; thus $x = 3$.

Since $f(3) = 0$, $g(0) = 3$.

$f'(x) = 3x^2 - 14x + 25$, $f'(3) = 3(3^2) - 14(3) + 25 = 10$

$g'(0) = \dfrac{1}{f'(3)} = \dfrac{1}{10}$

The correct choice is (C).

41. Since the variables cannot be separated, use Euler's method to approximate the solution:

    When $x = 1$ and $y = 3$, $\dfrac{dy}{dx} = 1(3) - 3^2 = -6$.

    Therefore, the approximate change in y corresponding to a change in x of one (from $x = 1$ to $x = 2$) is -6.

    Hence, $y(2) \approx 3 + (-6) = -3$.

    The correct choice is (A).

42. The area A of the rectangle may be represented by $A = 2x\sqrt{64 - x^2}$ where $2x$ is the base of the rectangle and its height is $y = \sqrt{64 - x^2}$.

    $$A' = (2x)\left(\dfrac{-x}{\sqrt{64 - x^2}}\right) + 2\sqrt{64 - x^2} = \dfrac{-2x^2 + 2(64 - x^2)}{\sqrt{64 - x^2}} = \dfrac{128 - 4x^2}{\sqrt{64 - x^2}}$$

    The critical values from the denominator are $x = \pm 8$. These are endpoint minimums; the height of the rectangle is zero.

    The other critical values occur when $128 - 4x^2 = 0 \Rightarrow x = \pm\sqrt{32}$.

    The derivative is negative when $x < -\sqrt{32}$, and $x > \sqrt{32}$. The derivative is positive when $-\sqrt{32} < x < \sqrt{32}$. Therefore the maximum occurs when $x = +\sqrt{32}$

    Therefore, the maximum area is $A = 2(\sqrt{32})(\sqrt{32}) = 64$.

    The correct choice is (E).

43. Use the method of integration by parts. $\int u\,dv = uv - \int v\,du$, with $u = x^n$, $du = nx^{n-1}$, $dv = \sin x\, dx$ and $v = -\cos x$. Then

    $$\int x^n \sin x\, dx = x^n(-\cos x) - \int (-\cos x)(nx^{n-1})\, dx$$
    $$= -x^n \cos x + n\int x^{n-1} \cos x\, dx$$

    The correct choice is (A).

44. Solve the differential equation by separating the variables and then using the method of partial fractions on the right side:

    $$\dfrac{dy}{y} = \dfrac{2\, dt}{t(t+2)}$$

$$\frac{A}{t} + \frac{B}{t+2} = \frac{A(t+2)+Bt}{t(t+2)} = \frac{2}{t(t+2)}$$

$$A(t+2) + Bt = 2$$

Substituting $t = -2 \Rightarrow B = -1$; substituting $t = 0 \Rightarrow A = 1$

$$\int \frac{dy}{y} = \int \frac{1}{t}\, dt + \int \frac{-1}{t+2}\, dt$$

$$\ln|y| = \ln|t| - \ln|t+2| + \ln C;$$

$$\ln|y| = \ln\left|\frac{Ct}{t+2}\right|;$$

$$e^{|y|} = e^{\left|\frac{Ct}{t+2}\right|}$$

$$y = C\left(\frac{t}{t+2}\right)$$

When $t = 1$, $y = 1$ (initial condition) $\Rightarrow 1 = C \cdot \frac{1}{3}$, or $C = 3$

The solution to the differential equation is $y = \frac{3t}{t+2}$.

Therefore, when $t = 2$, $y = \frac{3(2)}{2+2} = \frac{3}{2}$

The correct choice is (E).

45. The function f is a parabola with a maximum point at $(2,2)$. When $b = 2$ the horizontal line $y = 2$ is tangent at $(2,2)$. As b increases from 2, the tangent lines will move to the right of the maximum point. The "last" line tangent in the first quadrant will be just before the root of f at $x = 3$. Find the equation of the tangent line at $x = 3$. The slope is $f'(x) = -4(x-2)$ and $f'(3) = -4$. The tangent line at this point is $y = -4(x-3)$ or $y = -4x + 12$. Thus b must be between 2 and 12. (Note that $b \neq 12$ on the technicality that the axis is not in the first quadrant.)

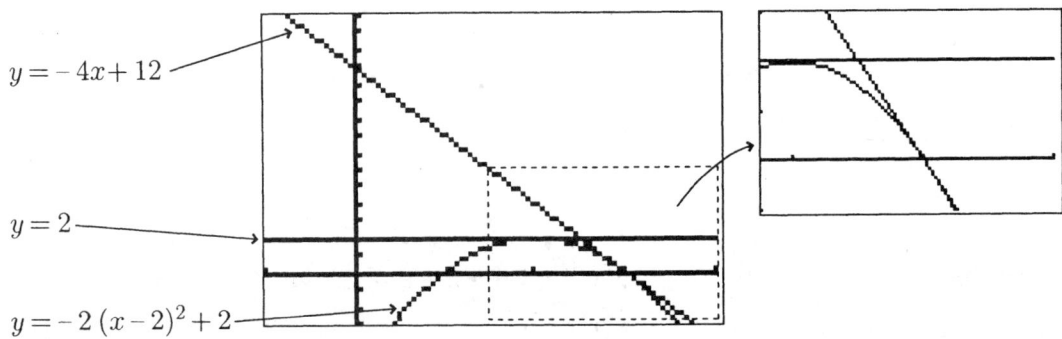

The correct choice is (C).

1a. The acceleration, $a(t)$, is the derivative of the velocity $v(t)$.

$$a(t) = v'(t) = -\frac{1}{t^2}\cos\left(\frac{1}{t}\right)$$

1b. The greatest velocity value occurs when $v'(t) = a(t) = 0$. This occurs when

$$\cos\left(\frac{1}{t}\right) = 0 \text{ or when } \frac{1}{t} = \frac{\pi}{2}, \frac{1}{t} = \frac{3\pi}{2}, \frac{1}{t} = \frac{5\pi}{2}, \text{ etc.}$$

$$t = \frac{2}{\pi}, \frac{2}{3\pi}, \frac{2}{5\pi}, \ldots$$

In the interval $0.1 < t < 0.3$, $v'(t) = a(t)$ is negative for $\frac{2}{5\pi} < t < \frac{2}{3\pi}$ and positive for $0.1 < t < \frac{2}{5\pi}$ and $\frac{2}{3\pi} < t < 0.3$. Therefore, the maximum occurs at $t = \frac{2}{5\pi} \approx 0.127$ or at the endpoint $t = 0.3$.

$$v\left(\frac{2}{5\pi}\right) = 1$$
$$v(0.3) \approx -0.191$$

The maximum occurs at $t = \frac{2}{5\pi}$

1c.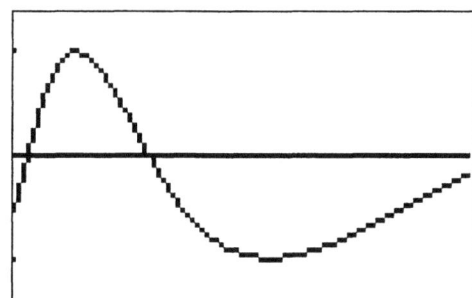

1d. The particle spends more time moving to the left. From the graph of $v(t)$, the velocity is below the x-axis (negative) much more than it is above, indicating that the particle moves more to the left than to the right.

2a. $\displaystyle\int_0^9 \frac{5000}{x^3+50}\,dx \approx 415.420$ or 415.421 cubic meters

2b. $S(6) - R(6) = \dfrac{5000}{6^3+50} - 23.967\sqrt{6} = -39.909$ or -39.910 cubic meters per hour. This means that the amount of sand in the bin is decreasing at the rate of 39.909 (or 39.910) cubic meters per hour when $t = 6$ hours.

2c. The amount of sand, $A(t)$, in the bin is given by $A(t) = \displaystyle\int_0^t S(x)\,dx - \int_1^t R(x)\,dx$. The maximum occurs when $A'(t) = S(t) - R(t) = 0$ or when $S(t) = R(t)$.

OR: Before the time when $S(t) = R(t)$ the rate at which sand is being poured into the bin is greater than the rate at which it is being taken out of the bin. After this point the rate at which it is taken out is greater than the rate at which it is pouring into the bin. Therefore the maximum must occur when $S(t) = R(t)$.

2d. From (a) the amount of sand put into the bin in the course of the workday is about 415.421 cubic meters. The amount that is taken out is $\displaystyle\int_1^9 23.967\sqrt{t}\,dt \approx 415.428$ cubic meters. Therefore, at the end of the workday the bin is empty.

3a. $r\dfrac{gallons}{year} \cdot 200 \; years = 200r \; gallons$

3b. Using the method of separation of variables:

$$\frac{dR}{R} = 0.05 dt$$
$$\ln|R| = 0.05t + C$$
$$R(t) = Ce^{0.05t}$$
$$R(0) = r = Ce^0 = C$$
$$R(t) = re^{0.05t}$$

3c. The amount of oil used is the definite integral of the rate of use. Therefore letting $x =$ the number of years until the oil supply is used up and $t = 0$ be the current year,

$$\int_0^x re^{0.05t} \, dt = 200r$$
$$r\left(\frac{e^{0.05t}}{0.05}\right)\bigg|_0^x = 200r$$
$$e^{0.05x} - 1 = 200(0.05) = 10$$
$$e^{0.05x} = 11$$
$$0.05x = \ln(11)$$
$$x = \frac{\ln(11)}{0.05} \approx 47.9579 \approx 48 \; years$$

Sample Examination III

4a. $\dfrac{dy}{dx} = \dfrac{dy/dt}{dx/dt} = \dfrac{6\sin^2 t \cos t}{-6\cos^2 t \sin t} = -\tan t$

4b. When $t = 0$, the coordinates are $(2, 0)$ which is point A. At this point, the slope is $\dfrac{dy}{dx} = -\tan 0 = 0$ which is the same as the slope of the x-axis.

When $t = \dfrac{\pi}{2}$, the coordinates are $(0, 2)$ which is point B. At this point, $\tan \dfrac{\pi}{2}$ is undefined as is the slope of the y-axis.

Thus, the slopes of the curved road and the two highways are the same at both A and B.

4c. The length of the curve is given by:

$$L = \int_a^b \sqrt{\left(\dfrac{dx}{dt}\right)^2 + \left(\dfrac{dy}{dt}\right)^2}\, dt = \int_0^{\frac{\pi}{2}} \sqrt{(-6\cos^2(t)\sin(t))^2 + (6\sin^2(t)\cos(t))^2}\, dt$$

5a. The series is geometric with a ratio of $-\frac{1}{3}(x-2)$. Therefore it will converge when

$$\left|-\tfrac{1}{3}(x-2)\right| < 1$$

$$-1 < \tfrac{1}{3}(x-2) < 1$$

$$-3 < x - 2 < 3$$

$$-1 < x < 5$$

5b. Differentiate term by term:

$$f'(x) = -\frac{1}{3} + \frac{2}{9}(x-2) - \frac{3}{27}(x-2)^2 + \frac{4}{81}(x-2)^3 + \ldots + (-1)^n(n)\frac{(x-2)^{n-1}}{3^n} + \ldots$$

Then substitute $x = 2$ to find $f'(2) = -\frac{1}{3}$.

5c. The tangent line has a slope of $-\frac{1}{3}$ and contains the point $(2, f(2)) = (2, 1)$. Its equation is $y - 1 = -\frac{1}{3}(x-2)$ or $y = 1 - \frac{1}{3}(x-2)$. Notice that this is the constant and linear term of the original power series. Alternatively, $f'(2)$ is the coefficient of the linear term of the power series; $f'(2) = -\frac{1}{3}$.

5d. Differentiating the answer to (b) gives

$$f''(x) = \frac{2}{9} - \frac{6}{27}(x-2) + \ldots$$

$$f''(2) = \frac{2}{9}$$

Since the second derivative is positive at $x = 2$, the graph is concave up there and the tangent line lies below the graph of f.

6a. $g(4.5) = 13.5$ (this is the area under the graph in the first quadrant between $x = 0$ and $x = 4.5$); $g'(4.5) = f(4.5) = 0$; $g''(4.5) =$ slope of f at $4.5 = -2$.

6b. Average value $= \dfrac{1}{5-(-3)} \displaystyle\int_{-3}^{5} f(x)\,dx = \dfrac{1}{8}\left[\int_{-3}^{3} f(x)\,dx + \int_{3}^{5} f(x)\,dx\right]$
$$= \frac{1}{8}[27 + 2] = \frac{29}{8}$$

The two integrals are found by finding the signed areas under the graph of f.

6c. There is a point of inflection where f, the derivative of g, has a maximum or minimum. This occurs only at $x = 5$.

6d. The function g increases from $x = -3$ to $x = 4.5$, then decreases until $x = 7$, and then increases to $x = 9$. There is a relative maximum at $(4.5, 13.5)$ (from part (a)). There is also an endpoint maximum at $x = 9$. Use the signed area to find $g(9)$:

$$g(9) = g(4.5) + \int_{4.5}^{9} f(x)\,dx$$
$$= 13.5 - 0.25$$

The endpoint maximum occurs at $(9, 13.25)$.

Thus the maximum points of g are $(4.5, 13.5)$ and $(9, 13.25)$.

Sample Examination IV

1. Treating y as an accumulation function, $\int_2^x f(t)\,dt$ gives the accumulated amount from $t=2$ to $t=x$. Adding to this the amount of 4 already present at $x=2$ gives $4+\int_2^x f(t)\,dt$.

   Alternately $y(x)-y(2) = \int_2^x f(t)\,dt$ so $y(x) = 4+\int_2^x f(t)\,dt$.

   The correct choice is (B).

2. The derivative of the expression is $24x^5 - 24x^2 = 24x^2(x^3-1) = 24x^2(x-1)(x^2+x+1)$. The derivative is equal to 0 when $x=0$ and $x=1$. (x^2+x+1 is always positive). Since the derivative changes from negative to positive only at $x=1$, this is the only relative minimum (First Derivative Test).

   $$\begin{array}{c|ccccc}
   f'(x) & - & 0 & - & 0 & + \\
   \hline
   x & & 0 & & 1 &
   \end{array}$$

   The correct choice is (C).

3. To find the vertical tangents, set $x'(t) = 0$.

   $x'(t) = -4\cos t$

   $x'(t) = 0$ when $-4\cos t = 0 \Rightarrow \cos t = 0$ or $t = \dfrac{\pi}{2}, \dfrac{3\pi}{2}$.

   Vertical tangents occur at all points of the form

   $\left(x\left(\dfrac{\pi}{2}\right), y\left(\dfrac{\pi}{2}\right)\right)$ and $\left(x\left(\dfrac{3\pi}{2}\right), y\left(\dfrac{3\pi}{2}\right)\right)$.

   So the vertical tangents are at $(-1, 4)$ and $(7, 4)$.

   Note: To find the horizontal tangents, set $y'(t) = 0$.

   The correct choice is (C).

4. $\dfrac{dx}{dt} = 4t$ and $\dfrac{dy}{dt} = 3t^2$

$$\dfrac{dy}{dx} = \dfrac{dy/dt}{dx/dt} = \dfrac{3t^2}{4t} = \dfrac{3t}{4}$$

$$\dfrac{d^2y}{dx^2} = \dfrac{\dfrac{d}{dt}\left(\dfrac{dy}{dx}\right)}{\dfrac{dx}{dt}} = \dfrac{3/4}{4t} = \dfrac{3}{16t}$$

Therefore at $t = 3$, $\dfrac{d^2y}{dx^2} = \dfrac{3}{16(3)} = \dfrac{1}{16}$.

The correct choice is (A).

5. From the answer choices it is apparent that what happens at $x = \pm\frac{1}{2}$ needs to be investigated.

Substitute $x = \dfrac{1}{2}$: $\displaystyle\sum_{n=1}^{\infty} \dfrac{2^n(\frac{1}{2})^n}{n} = \sum_{n=1}^{\infty} \dfrac{1}{n}$ this is the harmonic series. The harmonic series diverges. This eliminates choices (A), (B), and (D).

Substitute $x = -\dfrac{1}{2}$: $\displaystyle\sum_{n=1}^{\infty} \dfrac{2^n(-\frac{1}{2})^n}{n} = \sum_{n=1}^{\infty} \dfrac{(-1)^n}{n}$, this is the alternating harmonic series which converges. So the correct choice is (E) $-\frac{1}{2} \leq x < \frac{1}{2}$.

The correct choice is (E).

6. The average velocity of a particle for $t_1 \leq t \leq t_2$ is

$$\overline{v} = \dfrac{x(t_2) - x(t_1)}{t_2 - t_1}$$

where $x(t)$ is the position of the particle at time t.

The position function for this problem is given as $x(t) = \ln t$. Substituting into the expression above gives:

$$\overline{v} = \dfrac{\ln e - \ln 1}{e - 1} = \dfrac{1 - 0}{e - 1} = \dfrac{1}{e - 1}$$

The correct choice is (C).

7. The area is $\dfrac{1}{2}\displaystyle\int_\alpha^\beta r^2\,d\theta$ or equivalently $\dfrac{1}{2}\displaystyle\int_\alpha^\beta [f(\theta)]^2\,d\theta$.

The correct choice is (E).

8. Using the method of Integration by Parts with $u = 2x, du = 2dx, dv = \cos(2x)dx, v = \frac{1}{2}\sin(2x)$ gives

$$\int (2x)\cos(2x)dx = (2x)\left(\frac{1}{2}\sin(2x)\right) - \int \frac{1}{2}\sin(2x)(2dx)$$
$$= x\sin(2x) + \frac{1}{2}\cos(2x)$$

Or the set up could be $u = x, du = dx, dv = \cos(2x)(2dx), v = \sin(2x)$ which gives

$$\int (2x)\cos(2x)\,dx = x\sin(2x) - \int \sin(2x)\,dx$$
$$= x\sin(2x) + \frac{1}{2}\cos(2x)$$

The correct choice is (D).

9. Method I:

$$|x+2| = \begin{cases} x+2, & x \geq -2 \\ -x-2, & x < -2 \end{cases}$$

$$\int_{-3}^{3} |x+2|\,dx = \int_{-3}^{-2} (-x-2)\,dx + \int_{-2}^{3} (x+2)\,dx$$
$$= \left(\frac{-x^2}{2} - 2x\right)\Big|_{-3}^{-2} + \left(\frac{x^2}{2} + 2x\right)\Big|_{-2}^{3}$$
$$= \left[(-2+4) - (-\frac{9}{2}+6)\right] + \left[(\frac{9}{2}+6) - (2-4)\right]$$
$$= (2 + \frac{9}{2} - 6) + (\frac{9}{2} + 6 - 2 + 4)$$
$$= 13$$

Method II:

Graph the function and find the areas of the two triangles.

$$\frac{1}{2}(1)(1) + \frac{1}{2}(5)(5) = 13$$

The correct choice is (C).

10.

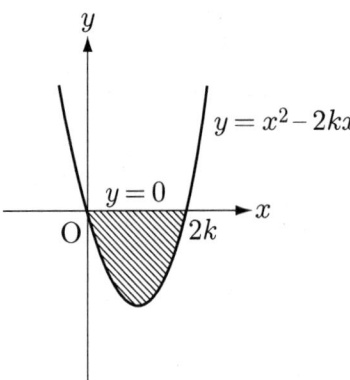

The graph of $y = x^2 - 2kx$ is a parabola whose x-intercepts are $x = 0$ and $x = 2k$.

The x-intercepts are found by setting $y = 0$ and solving for x.

$$x^2 - 2kx = 0$$

$$x(x - 2k) = 0$$

$$x = 0, x = 2k$$

The region R is shown shaded in the figure above. <u>Note</u>: The area of region R is bounded by two functions, $y_1 = 0$ and $y_2 = x^2 - 2kx$. Between $x = 0$ and $x = 2k$, $y_1 > y_2$. Since the area of the region is given as 36,

$$\int_0^{2k} 0 - (x^2 - 2kx)\, dx = 36$$

$$\int_0^{2k} (-x^2 + 2kx)\, dx = 36$$

$$\left. \frac{-x^3}{3} + \frac{2kx^2}{2} \right|_0^{2k} = 36$$

$$\left. \frac{-x^3}{3} + kx^2 \right|_0^{2k} = 36$$

$$\frac{-8k^3}{3} + 4k^3 = 36$$

$$4k^3 = 108$$

$$k^3 = 27$$

$$k = 3$$

The correct choice is (B).

Sample Examination IV   63

11. The given problem is equivalent to $\lim_{h \to 0} \dfrac{\int_0^h \frac{\sin^2 t}{t^2} dt}{h}$ which is of the form $\dfrac{0}{0}$.

   Using L'Hôpital's Rule, take the first derivatives of the numerator (by the Fundamental Theorem of Calculus) and denominator,

   $$\lim_{h \to 0} \dfrac{\int_0^h \frac{\sin^2 t}{t^2} dt}{h} = \lim_{h \to 0} \dfrac{\frac{\sin^2 h}{h^2}}{1}$$
   $$= \lim_{h \to 0} \dfrac{\sin^2 h}{h^2} = \lim_{h \to 0} \left(\dfrac{\sin h}{h} \cdot \dfrac{\sin h}{h} \right) = 1$$

   Keep in mind that $\lim_{h \to 0} \dfrac{\sin h}{h} = 1$.

   The correct choice is (C).

12. Since $f(x)$ is continuous on $[-3, 7]$ and differentiable on $(-3, 7)$, the Mean Value Theorem guarantees a value c, $-3 < c < 7$, such that $f'(c) = \dfrac{f(7) - f(-3)}{7 - (-3)} = \dfrac{2 - 4}{10} = -\dfrac{1}{5}$

   The correct choice is (B).

13. The easiest method is to realize that $\dfrac{1 - x}{x - 1} = -1$. Therefore, the derivative of -1 is 0.

   The Quotient Rule could also be used to get the same result.

   The correct choice is (B).

14. Write $g(x)$ as $g(x) = [f(x)]^{-1}$, then differentiate using the power rule and the chain rule.

   $g'(x) = (-1)[f(x)]^{-2} f'(x)$. Then substitute $x = 2$ and use the values from the table,

   $$g'(2) = (-1)[f(2)]^{-2} f'(2) = (-1)(-8)^{-2}(-4) = \dfrac{4}{64} = \dfrac{1}{16}$$

   Note: The Quotient Rule may also be used to find $g'(x)$.

   $$g'(x) = \dfrac{-f'(x)}{[f(x)]^2}; \text{ so } g'(2) = \dfrac{-f'(2)}{[f(2)]^2} = \dfrac{4}{64} = \dfrac{1}{16}$$

   The correct choice is (C).

15. I. $\sum_{n=1}^{\infty}\left(1-\frac{4}{3}\right)^n$ is a geometric series whose common ratio is $\left(-\frac{1}{3}\right)$ and will converge.

    II. $\sum_{n=1}^{\infty}\left(1+\frac{4}{3}\right)^n$ is also a geometric series whose ratio is $\frac{7}{3}$ and will diverge.

    For III, $\lim_{n\to\infty}\left(1+\frac{1}{n}\right)^n = e$ and therefore it will diverge by the nth term test.

    The correct choice is (A).

16. This could be done by differentiating the given expression three times. However, the general term of the Taylor Series is $\dfrac{f^{(n)}(a)(x-a)^n}{n!}$.

    The third derivative ($n=3$) at $x=a$ is found in the fourth term; it is the coefficient of $\dfrac{(x-3)^3}{3!}$. Thus $f'''(3) = 7$.

    The correct choice is (E).

17. Substituting for $f(x)$ gives:

    $$\int_{-2}^{2}(f(x)-g(x))\,dx = \int_{-2}^{2}(15-g(x)-g(x))\,dx$$
    $$= \int_{-2}^{2}(15-2g(x))\,dx$$
    $$= 15\int_{-2}^{2}dx - 2\int_{-2}^{2}g(x)\,dx$$
    $$= 15x\Big|_{-2}^{2} - 2\int_{-2}^{2}g(x)\,dx$$
    $$= 60 - 2\int_{-2}^{2}g(x)\,dx$$

    Note: Since there is no information about symmetry, it cannot be assumed that $\int_{-2}^{2}g(x)\,dx = 2\int_{0}^{2}g(x)\,dx$. Thus choice (D) is not correct.

    The correct choice is (E).

18. $\int_a^x \dfrac{d}{dt}(f(t))\, dt = f(t)\Big|_a^x = f(x) - f(a)$

So, $\int_2^4 \left[\dfrac{d}{dt}(3t^2 + 2t - 1)\right] dt = 3t^2 + 2t - 1 \Big|_2^4 = (48 + 8 - 1) - (12 + 4 - 1) = 55 - 15 = 40$

Note: Do not confuse $\int_a^x \dfrac{d}{dt}[f(t)]\, dt$ with $\dfrac{d}{dx}\left[\int_a^x f(t)\, dt\right]$.

$\int_a^x \dfrac{d}{dt}(f(t))\, dt = f(x) - f(a)$, while $\dfrac{d}{dx}\left[\int_a^x f(t)\, dt\right] = f(x)$.

The correct choice is (B).

19. The comparison test shows that I is true since $a_n < b_n$. II could be true, but there is no convincing reason to believe it is. III is true since the sum of two convergent series converges.

The correct choice is (D).

20. Since each part of the definition of the function is continuous, check the place where they join, at $x = 5$. Here, both parts reduce to $1 + e^{-5}$. Therefore, the function is continuous, so I is true.

$f'(x) = \begin{cases} -e^{-x}, & 0 \leq x \leq 5 \\ e^{x-10}, & 5 < x \leq 10 \end{cases}$

The derivative on the left side of 5 is negative and on the right side is positive. It is never zero so the derivative is not continuous, and II is false.

$f''(x) = \begin{cases} e^{-x} & 0 \leq x \leq 5 \\ e^{x-10}, & 5 < x \leq 10 \end{cases}$

The second derivative is always positive, so the graph of $f(x)$ is concave up everywhere, and III is true.

The correct choice is (D).

21. The equation of the circle is $x^2 + y^2 = 9$.

Recall that for an equilateral triangle $A = \dfrac{\sqrt{3}}{4}S^2$. The cross section of the solid is an equilateral triangle of side $2y$ and area $A(x) = \sqrt{3}y^2$ or $\sqrt{3}(9 - x^2)$.

The volume $V = \sqrt{3} \int_{-3}^{3} (9 - x^2) \, dx$

$= \sqrt{3} \left(9x - \dfrac{x^3}{3} \right) \Big|_{-3}^{3}$

$= \sqrt{3}(27 - 9) - \sqrt{3}(-27 + 9)$

$= 36\sqrt{3}$

The correct choice is (E).

22. Using the substitution $u = 25 - x^2 \Rightarrow du = -2x \, dx$ or $x \, dx = -\tfrac{1}{2} \, du$,

then $\int x\sqrt{25 - x^2} \, dx = -\dfrac{1}{2} \int \sqrt{u} \, du$.

Also change the limits of integration to u-limits; when $x = 0$, $u = 25 - 0^2 = 25$ and when $x = 3$, $u = 16$.

Therefore, $\int_0^3 x\sqrt{25 - x^2} \, dx = -\tfrac{1}{2} \int_{25}^{16} \sqrt{u} \, du$.

Note that none of the given choices has the direct expression of $-\dfrac{1}{2} \int_{25}^{16} \sqrt{u} \, du$.

However, keep in mind that $\int_a^b f(x) \, dx = -\int_b^a f(x) \, dx$.

So, $-\dfrac{1}{2} \int_{25}^{16} \sqrt{u} \, du = \dfrac{1}{2} \int_{16}^{25} \sqrt{u} \, du$.

The correct choice is (D).

23. Cancelling the 8^n gives $s_n = \left(\dfrac{1}{8^2}\right) \left(\dfrac{(8-n)^{200}}{(3-n^2)^{100}}\right)$.

If the binomials in the second fraction are expanded, both numerator and denominator will be polynomials with leading (highest power) terms of $(+1)n^{200}$. The limit of this part of the expression is $(+1)$, so the limit of the entire expression is $\tfrac{1}{64}$.

The correct choice is (D).

24. "Increasing" indicates the function rises as x increases. "At a decreasing rate" indicates that the slope gets less steep as x increases. Only the graph of (C) is increasing at a decreasing rate.

The correct choice is (C).

25. This question may be answered without solving the differential equation. For horizontal asymptotes $\lim_{x \to \pm\infty} \frac{dy}{dx} = 0$. The derivative will be zero when $y = 0$ and when $\left(8 - \frac{y}{1,000}\right) = 0$ or $y = 8,000$.

Or solving gives $y = \frac{8000e^{8x}}{e^{8x} + C}$. $\lim_{x \to +\infty} \frac{8000e^{8x}}{e^{8x} + C} = 8,000$ and $\lim_{x \to -\infty} \frac{8000e^{8x}}{e^{8x} + C} = 0$

The correct choice is (E).

26. The given limit is actually the definition of the derivative of $\tan 2x$.

If $f(x) = \tan 2x$, $f'(x) = \lim_{h \to 0} \frac{f(x+h) - f(x)}{h}$
$= \lim_{h \to 0} \frac{\tan(2(x+h)) - \tan(2x)}{h}$

Consequently, the derivative of $\tan 2x$ is $2\sec^2(2x)$.

The correct choice is (B).

27. The slope field gives the slope of the solution of the differential equation. This function increases, levels off, and then increases again. So, I is false; and, II and III are true.

The correct choice is (D).

28. At point A, the function is decreasing indicating $f' < 0$ and concave up indicating that $f'' > 0$, thus $f' < f''$ is true at A.

At point B, the function is increasing indicating $f' > 0$ and concave down indicating that $f'' < 0$, thus $f' < f''$ is false at B.

At point C, the function has a horizontal tangent indicating $f' = 0$ and concave down indicating that $f'' < 0$, thus $f' < f''$ is false at point C.

Hence, $f'(x) < f''(x)$ at point A only.

The correct choice is (A).

29. __Method I:__

To find $\dfrac{dy}{dx}$, use implicit differentiation,

$$e^{xy} = 2$$
$$(e^{xy})\left(x\dfrac{dy}{dx} + y\right) = 0$$
$$xe^{xy}\dfrac{dy}{dx} + ye^{xy} = 0$$
$$\dfrac{dy}{dx} = \dfrac{-ye^{xy}}{xe^{xy}}\bigg|_{(1,\ln 2)} = \dfrac{(-\ln 2)(e^{\ln 2})}{e^{\ln 2}} = -\ln 2$$

__Method II:__

Alternate Solution:

$$\ln e^{xy} = \ln 2$$
$$xy = \ln 2$$
$$y = \dfrac{\ln 2}{x}$$
$$\dfrac{dy}{dx} = \dfrac{-\ln 2}{x^2}$$
$$\dfrac{dy}{dx}\bigg|_{x=1} = -\ln 2$$

The correct choice is (A).

30. Speed $= |x'(t)| = |4t^3 - 30t^2 + 58t - 36|$. The greatest speed may be found by using your calculator to graph, or otherwise evaluate, $|4t^3 - 30t^2 + 58t - 36|$. Of the values given, $t = 4$ gives the greatest speed. (Note: the greatest speed in the interval is at $t \approx 3.690$).

The correct choice is (D).

31. You are given the graph of $f'(x)$ and you must obtain the graph of $f(x)$. First, observe that since $f'(x) > 0$ for all x, $f(x)$ is increasing. Thus eliminate the graph of choices (A), (B), and (C). When the derivative is increasing ($x \leq 0$) the second derivative will be positive and the function will be concave upwards. Likewise, when the derivative is decreasing ($x \geq 0$) the function will be concave downward. This describes graph (E) and eliminates graph (D).

The correct choice is (E).

32. Integrate the velocity vector to find the position vector: $\vec{s}(t) = (-2\cos t + C_1, 3\sin t + C_2)$.

Substitute $t = 0$ to evaluate the two constants, C_1 and C_2:

$$-2\cos(0) + C_1 = 1 \Rightarrow C_1 = 3 \text{ and } 3\sin(0) + C_2 = 1 \Rightarrow C_2 = 1$$

So the position vector is $(-2\cos t + 3, 3\sin t + 1)$. Substitute $t = 2$ to find the position vector:

$$(-2\cos(2) + 3, 3\sin(2) + 1) = (3.832, 3.728).$$

The correct choice is (A).

33. With $f(x) = 2 - \cos x$ the approximation is
$$\frac{1}{2}(4)(f(1) + f(5)) + \frac{1}{2}(3)(f(5) + f(8)) + \frac{1}{2}(2)(f(8) + f(10))$$

Notice that each of the above terms is the formula for the area of the separate trapezoids. Substituting the values gives
$$\frac{1}{2}(4)(1.4597 + 1.7163) + \frac{1}{2}(3)(1.7163 + 2.1455) + \frac{1}{2}(2)(2.1455 + 2.8391) \approx 17.129$$

Note the uneven subintervals prevent you from using a Trapezoidal Rule program that you may have in your calculator. Such a program, with 3 equal subintervals, gives the value 18.147 (choice D).

Also note that in doing any computation you must use at least one more digit than the final answer requires. In this case the computation, done on your calculator, needs at least 4 decimal places. Round off LAST. Rounding or truncating too soon gives a close, but incorrect answer. Finally, integration by using a calculator gives the value 19.385 (choice E).

The correct choice is (B).

34. $y' = 3x^2$; Arc length $= \int_a^b \sqrt{1 + \left(\frac{dy}{dx}\right)^2}\, dx = \int_0^1 \sqrt{1 + 9x^4}\, dx = 1.548$

The correct choice is (D).

35. Graph $f'(x)$ in the viewing window $x[0, 12]$ by $y[-2, 2]$. In the given interval, the derivative changes from positive to negative twice indicating that $f(x)$ will have two relative maxima (First Derivative Test).

The correct choice is (C).

36. Method I:

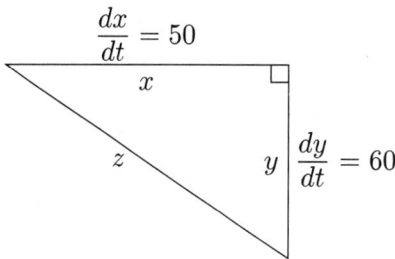

Let x be the distance traveled by the first car and let y be the distance traveled by the second car.

You know that $dx/dt = 50$ and $dy/dt = 60$ and the distance between them is $z = \sqrt{x^2 + y^2}$. (see above diagram)

You want to find dz/dt (the rate at which the distance between them is changing) when $x = 25$ and $y = 30$ (one-half hour later).

Taking the derivative of both sides with respect to t, you get

$$\frac{dz}{dt} = \frac{2x\frac{dx}{dt} + 2y\frac{dy}{dt}}{2\sqrt{x^2 + y^2}}$$

One-half hour after they start $x = 25$ and $y = 30$. Substitute those values (and the values of dx/dt and dy/dt) to get (with your calculator) $dz/dt = 78.102$ or about 78 miles per hour.

Method II:

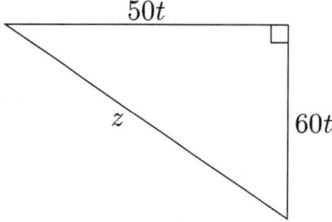

Express the distance traveled as a function of time t. The first car will travel $50t$ miles and the second car will travel $60t$ miles. The distance between them is given by

$$z = \sqrt{(50t)^2 + (60t)^2} \quad \text{and} \quad \frac{dz}{dt} = \frac{2(50t)(50) + 2(60t)(60)}{2\sqrt{(50t)^2 + (60t)^2}}$$

Substitute $t = \frac{1}{2}$ (one-half hour later) and find that $dz/dt \approx 78$ miles per hour. (Recall that $x = 50t$, $y = 60t$, $dx/dt = 50$, and $dy/dt = 60$).

The correct choice is (D).

37. The area of the square is 4. The area under the parabola $y = -x^2 + 2x$ is given by

$$\int_0^2 (-x^2 + 2x)\, dx = -\frac{x^3}{3} + x^2 \Big|_0^2 = \frac{4}{3}$$

The area above the parabola is $4 - \left(\frac{4}{3}\right) = \frac{8}{3}$.

The probability is the ratio of the area above the parabola to the total area: $\frac{8/3}{4} = \frac{2}{3}$.

The correct choice is (C).

38. The second derivative, $f''(x) = 5(x^2 + 5)^4(2x)$, changes sign only once, at $x = 0$. The other factors are all positive for all values of x. Therefore, there is only one point of inflection.
The correct choice is (B).

39. Begin by finding the antiderivative term by term. Note the constant of integration:

$$f(x) = \int \left(3x - \frac{9x^3}{2} + \frac{81x^5}{40} - \frac{3x^7}{60} + \cdots\right) dx$$

$$= C + \frac{3x^2}{2} - \frac{9x^4}{2 \cdot 4} + \frac{81x^6}{40 \cdot 6} - \frac{3x^8}{60 \cdot 8} + \cdots$$

$$= C + \frac{3x^2}{2} - \frac{9x^4}{8} + \frac{27x^6}{80} - \frac{3x^8}{480} + \cdots$$

Using the initial condition $f(0) = 2$ and find C:

$$2 = C + \frac{3(0)^2}{2} - \frac{9(0)^4}{8} + \frac{27(0)^6}{80} - \frac{3(0)^8}{480} + \cdots$$

Therefore, $C = 2$ and

$$f(x) = 2 + \frac{3x^2}{2} - \frac{9x^4}{8} + \frac{27x^6}{80} - \frac{3x^8}{480} + \cdots$$

The correct choice is (E).

40. Analytically, the problem can be solved by finding which point's x-coordinate reaches 6π first. To do this have your calculator solve $x_\alpha = (3t - 3\sin t) = 6\pi$ and $x_\beta = (3t - 4\sin t) = 6\pi$. $x_\alpha = 6\pi$ when $t = 6.283$ and $x_\beta = 6\pi$ when $t = 5.007$. Thus, Beta arrives first, Alpha loses. It is probably more fun to watch the race: Enter the equations for Alpha and Beta in the calculator's parametric graphing mode. Also enter the finish lines at $x(t) = 6\pi, y(t) = t$. Graph simultaneously and note that Beta starts in the wrong direction, Alpha's path is shorter, but Beta moves faster and Beta wins. The photo finish:

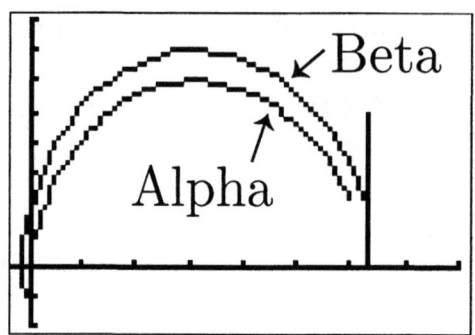

The correct choice is (A).

41. Use the method of separation of variables to solve the differential equation:

$$\frac{dy}{y} = e^x \, dx$$
$$\ln|y| = e^x + C_1$$
$$|y| = e^{e^x + C_1}$$
$$y = Ce^{e^x}$$

Then substitute the initial condition ($y = 3$ when $x = 0$) to find C:

$$3 = Ce^{e^0}$$
$$3 = Ce$$
$$\frac{3}{e} = C$$

Therefore, $y = \frac{3}{e} e^{e^x} = \frac{3e^{e^x}}{e}$

The correct choice is (E).

42. Use implicit differentiation:

$$2y\frac{dy}{dx} - 3x^2 - 30x = 0$$
$$\frac{dy}{dx} = \frac{3x^2 + 30x}{2y}$$

The tangent lines will be horizontal when

$$\frac{3x^2 + 30x}{2y} = 0$$

$$3x(x + 10) = 0$$

$$x = 0, -10$$

The correct choice is (D).

43. __Method I__:

   Solve the differential equation by separating the variables:
   $$\frac{dy}{y} = k\, dt$$
   $$\ln |y| = kt + C_1$$
   $$|y| = e^{kt+C_1}$$
   $$y = Ce^{kt} \text{ where } C = e^{C_1} > 0$$

   When $t = 0$ then $y = y(0)$
   $$y(0) = Ce^0$$
   $$y(0) = C$$

   Substitute the given condition that $y(20) = \frac{1}{2}y(0)$ and solve for k:
   $$\frac{1}{2}y(0) = y(0)e^{20k}$$
   $$\frac{1}{2} = e^{20k}$$
   $$\ln\left(\frac{1}{2}\right) = 20k$$
   $$\frac{\ln\left(\frac{1}{2}\right)}{20} = k$$
   $$-0.035 = k$$

   __Method II__:

   The equation $y(t) = y(0)e^{-\frac{\ln 2}{T}t}$, where T is the half-life, models this situation. For a half-life of 20 days, $k = -\frac{\ln 2}{20} = -0.035$.

   The correct choice is (D).

44. a is a constant equal to $f(2)$ which is negative, as seen on the graph.

   b will correspond to $f'(2) = 0$, since at $x = 2$, f has a relative minimum.

   c corresponds to $\frac{f''(2)}{2!}$ which is greater than 0, since $f(x)$ is concave upward at $x = 2$.

   __Alternate Solution__: The Taylor polynomial is a parabola whose vertex is at the point $(2, a)$ which is also the relative minimum point of the function graphed. Therefore $a < 0$, $b = 0$. Since the curve is concave upward at this point, the parabola must open upward, therefore $c > 0$.

   The correct choice is (A).

45. Euler's Method, the tangent line approximation, calculates the approximate value of y as

$$y \approx f(x) + f'(3)\Delta x = -2 + \left(-\frac{3^2}{-2}(-0.3)\right) = -2 - 1.35 = -3.35$$

The correct choice is (E).

1a. $\lim\limits_{x \to -\infty} f(x) = 0$; $\lim\limits_{x \to \infty} f(x) = 0$. As x increases or decreases without bound, $e^{-x^2/2}$ approaches zero.

1b. Copy the graph from your calculator.

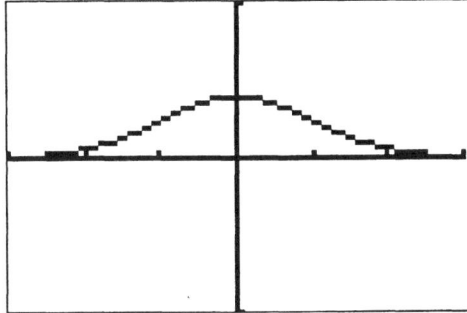

1c. $f'(x) = \dfrac{1}{\sqrt{2\pi}} \left(e^{-\frac{x}{2}^2} \right)(-x)$

$f''(x) = \dfrac{1}{\sqrt{2\pi}} \left(-e^{-\frac{x}{2}^2} + x^2 e^{-\frac{x}{2}^2} \right)$

$ = \dfrac{1}{\sqrt{2\pi}} \left(e^{-\frac{x}{2}^2} \right)(x^2 - 1)$

The second derivative changes sign at $x = 1$ and $x = -1$. These are the x-coordinates of the points of inflection.

1d. The area is $\displaystyle\int_{-1}^{1} \dfrac{1}{\sqrt{2\pi}} e^{-\frac{x}{2}^2} \, dx = 0.682$ or 0.683, by use of a graphing calculator.

2a.

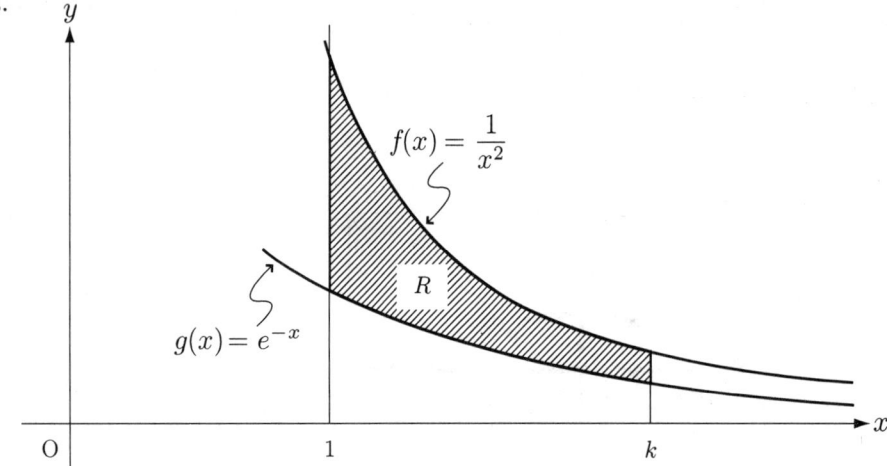

2b. $A(k) = \int_1^k \left[\dfrac{1}{x^2} - e^{-x}\right]\,dx = \left(-\dfrac{1}{x} + e^{-x}\right)\Big|_1^k = -\dfrac{1}{k} + \dfrac{1}{e^k} + 1 - \dfrac{1}{e}$

2c. $\lim\limits_{k\to\infty} A(k) = \lim\limits_{k\to\infty}\left(-\dfrac{1}{k} + \dfrac{1}{e^k} + 1 - \dfrac{1}{e}\right) = -0 + 0 + 1 - \dfrac{1}{e} = 1 - \dfrac{1}{e}$

2d. By the Disk Method: $V_x = \pi \int_1^4 \left[\left(\dfrac{1}{x^2}\right)^2 - (e^{-x})^2\right]\,dx \approx 0.260\pi$, or 0.261π, or 0.818, or 0.819

3a. Substitute $L = 10,000$ when $d = 40$ into $L = \dfrac{k}{d^2}$ to find k:

$$10,000 = \frac{k}{40^2} \Rightarrow k = 10^4 \cdot 40^2 \text{ or } 1.6 \times 10^7$$

3b. From trigonometry, $\dfrac{40}{d} = \cos\theta$, so $d = \dfrac{40}{\cos\theta}$

Since $L = \dfrac{k}{d^2}$, $L(\theta) = \dfrac{(1.6)(10^7)}{\left(\frac{40}{\cos\theta}\right)^2} = 10^4 \cos^2\theta$

3c. Differentiate $L = 10^4 \cos^2\theta$ (from part (b)) with respect to t (time) to find the rate of change of L:

$$\frac{dL}{dt} = 2 \cdot 10^4 (\cos\theta)(-\sin\theta)\frac{d\theta}{dt}$$

Then substitute $\theta = \dfrac{\pi}{4}$ and $\dfrac{d\theta}{dt} = \dfrac{\pi}{30}$ (from the given) to find the numerical value:

$$\frac{dL}{dt} = 2 \cdot 10^4 \cos\left(\frac{\pi}{4}\right)\left(-\sin\frac{\pi}{4}\right)\left(\frac{\pi}{30}\right)$$

$$= 2 \cdot 10^4 \left(\frac{\sqrt{2}}{2}\right)\left(-\frac{\sqrt{2}}{2}\right)\left(\frac{\pi}{30}\right)$$

$$= \frac{-10^4 \pi}{30} = -1047.198$$

Be careful how you phrase your answer: "The strength is decreasing at $+1047.198$ lumens/sec." Or, "The strength is changing at -1047.198 lumens/sec". The negative sign indicates that it is decreasing. Do NOT write "The strength is decreasing at -1047.198 lumens/sec."

3d. Solving $10^4 \cos^2\theta = 1,000 \Rightarrow \cos^2\theta = \dfrac{1,000}{10,000} \Rightarrow \cos^2\theta = \dfrac{1}{10} \Rightarrow \theta = 1.249$ radians.

Since L starts at more than 1,000 lumens, after this value $L < 1,000$ lumens. Therefore, the value of θ is 1.249 radians.

4a. To solve the differential equation, first separate the variables:

$$\frac{dh}{dt} = -k\sqrt{h}$$

$$\frac{dh}{\sqrt{h}} = -k \, dt$$

Integrating both sides,

$$2\sqrt{h} = -kt + C$$

Now, use the initial condition $h(0) = 16$ to find C.

$$2\sqrt{16} = -k(0) + C \Rightarrow C = 8$$

$$2\sqrt{h} = -kt + 8.$$

Now square both sides and solve for h:

$$4 \cdot h(t) = (-kt + 8)^2$$

$$h(t) = \tfrac{1}{4}(8 - kt)^2$$

4b. When $t = 8$, $h = 12.25$;

So, $12.25 = \tfrac{1}{4}(8 - 8k)^2$

Multiply both sides by 4 and take the positive square root (Note that using the negative square root gives a value of $k > 1$):

$7 = 8 - 8k \Rightarrow k = \tfrac{1}{8} = 0.125$

4c. The tank will be completely empty when $h(t) = 0$.

$0 = \tfrac{1}{4}(8 - \tfrac{1}{8}t)^2$

Solving for t, the result is $t = 64$ hours.

After 64 hours, the tank will be completely empty.

5a. To find the point(s), first find the slope of chord AB and set it equal to the derivative of f:

Slope of chord $AB = \dfrac{15-(-15)}{-3-3} = -5$

$f'(x) = 4 - 3x^2$

So, $4 - 3x^2 = -5$

$3x^2 = 9$

$x^2 = 3$

$x = \pm\sqrt{3}$

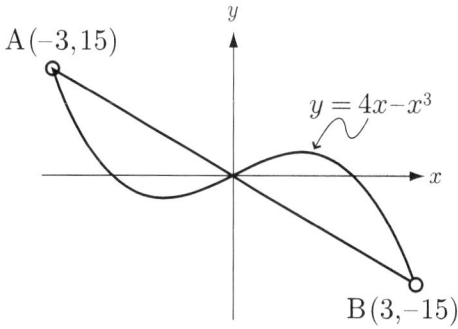

Substitute $x = \sqrt{3}$ and $x = -\sqrt{3}$ into the curve $y = 4x - x^3$ to get $(\sqrt{3}, \sqrt{3})$ and $(-\sqrt{3}, -\sqrt{3})$.

Looking at the diagram, you can see there are 2 points on the curve where the tangent line is parallel to chord AB.

5b. For $0 < x < 3$, $f(x)$ lies above the chord AB (see diagram above), so the vertical distance is $V(x) = (4x - x^3) - (-5x) = 9x - x^3$.

Note: Chord AB with coordinates $(-3,, 15)$ and $(3, -15)$ has an equation of $y = -5x$.

5c. To find the maximum vertical distance, differentiate $V(x)$ and find its roots:

$V(x) = 9x - x^3$

$V'(x) = 9 - 3x^2$; $V'(x) = 0$ when $9 - 3x^2 = 0 \Rightarrow 3x^2 = 9 \Rightarrow x^2 = 3$ or $x = \sqrt{3}$

(Note: $x = -\sqrt{3}$ is not in the domain $0 < x < 3$)

$x = \sqrt{3}$ is a maximum since $V''(x) = -6x < 0$ at $x = \sqrt{3}$ (Second Derivative Test)

So the maximum vertical distance is $V(\sqrt{3}) = 9\sqrt{3} - (\sqrt{3})^3 = 6\sqrt{3}$

Note: You must substitute $x = \sqrt{3}$ into $V(x)$ to find the maximum vertical distance, and not leave the answer as $x = \sqrt{3}$.

6a. The slope is $f'(\frac{\pi}{2}) = \ln(\sin\frac{\pi}{2}) = \ln(1) = 0$. The point is $(\frac{\pi}{2}, 1)$ and an equation of the line is $y - 1 = 0(x - \frac{\pi}{2})$ or $y = 1$.

6b. Use the Chain Rule to differentiate f': $f''(x) = \dfrac{1}{\sin x} \cos x = \cot x$

6c. Yes, the function changes concavity at $x = \frac{\pi}{2}$. To the left of this value $\cot x > 0$ and to the right $\cot x < 0$. Since the second derivative changes sign, you can conclude that the function changes concavity there.

6d. From the given information and the results of the previous parts you know that $f(\frac{\pi}{2}) = 1$, $f'(\frac{\pi}{2}) = 0$, and $f''(\frac{\pi}{2}) = \cot\frac{\pi}{2} = 0$. Then $f'''(x) = -\csc^2 x$ and $f'''(\frac{\pi}{2}) = -1$.

So the Taylor polynomial of degree three is

$$T_3(x) = 1 + \frac{0}{1!}\left(x - \frac{\pi}{2}\right) + \frac{0}{2!}\left(x - \frac{\pi}{2}\right)^2 - \frac{1}{3!}\left(x - \frac{\pi}{2}\right)^3$$
$$= 1 - \frac{1}{6}\left(x - \frac{\pi}{2}\right)^3$$

Sample Examination V

1. If $y = (2x^2 + 1)^4$, then $\dfrac{dy}{dx} = 4(2x^2 + 1)^3(4x) = 16x(2x^2 + 1)^3$.

   The correct choice is (E).

2. Let $u = x^2 + 1$

   $du = 2x \, dx \Rightarrow x \, dx = \dfrac{1}{2} \, du$

   $\displaystyle\int x(x^2 + 1)^{\frac{1}{2}} \, dx = \tfrac{1}{2} \int u^{\frac{1}{2}} \, du = \tfrac{1}{2}(\tfrac{2}{3} u^{\frac{3}{2}}) = \tfrac{1}{3} u^{\frac{3}{2}} = \tfrac{1}{3}(x^2 + 1)^{\frac{3}{2}} + C$

   The correct choice is (C).

3. This question is testing a consequence of the Intermediate Value Theorem, i.e., if f is continuous on $[a, b]$ and $f(a)$ and $f(b)$ differ in sign, then the equation $f(x) = 0$ has at least one solution in the open interval (a, b).

   In the example, $f(x) = x^3 - x + 3$ and $f(c) = 0$ for only one real number c.

   Examine the choices: does $f(c) = 0$ when c is between -2 and -1?

   Well, $f(-2) = -3 < 0$ and $f(-1) = 3 > 0$, therefore c must be between -2 and -1.

   The correct choice is (A).

4. Since $x = 2t + 3$ and $y = t^2 + 2t$, $\dfrac{dy}{dt} = 2t + 2$ and $\dfrac{dx}{dt} = 2$.

   So $\dfrac{dy}{dx} = \dfrac{dy/dt}{dx/dt} = \dfrac{2t + 2}{2} = \dfrac{2(t + 1)}{2} = t + 1$.

   At $t = 1$, $\dfrac{dy}{dx} = 2$. When $t = 1$, $x = 5$ and $y = 3$.

   So the tangent line to the curve at $t = 1$ is $y - 3 = 2(x - 5)$ or $y = 2x - 7$.

   The correct choice is (A).

5. $\int_0^8 \frac{1}{\sqrt[3]{8-x}} dx$ is an improper integral.

$$\int_0^8 \frac{1}{\sqrt[3]{8-x}} dx = \lim_{b \to 8^-} \int_0^b \frac{1}{\sqrt[3]{8-x}} dx = \lim_{b \to 8^-} \left[-\frac{3}{2}(8-x)^{\frac{2}{3}} \right]_0^b = -\frac{3}{2}(0) + \frac{3}{2}(4) = 6$$

The correct choice is (C).

6. This problem is solved using integration by parts.
$\quad u = x \quad\quad dv = \sin x\, dx$
$\quad du = dx \quad\quad v = -\cos x$

$$\int x \sin x\, dx = uv - \int v\, du = -x \cos x - \int -\cos x\, dx$$
$$= -x \cos x + \int \cos x\, dx$$
$$= -x \cos x + \sin x + C$$

The correct choice is (D).

7. Statement I is false. Since $\int_0^3 f(x)\, dx$ is constant, $\frac{d}{dx} \int_0^3 f(x)\, dx = 0$ (by the Fundamental Theorem of Calculus).

Statement II is false because $\int_3^x f'(x)\, dx = f(x) - f(3)$ (by the Fundamental Theorem of Calculus).

Statement III is true because of the Fundamental Theorem of Calculus.

The correct choice is (B).

8. To find $\frac{dy}{dx}$, use implicit differentiation.

$$\sin(xy) = x^2$$
$$(xy' + y) \cos(xy) = 2x$$
$$xy' \cos(xy) + y \cos(xy) = 2x$$
$$xy' \cos(xy) = 2x - y \cos(xy)$$
$$y' = \frac{2x - y \cos(xy)}{x \cos(xy)}$$

Now, since $\frac{1}{\cos(xy)} = \sec(xy)$, $y' = \frac{2x \sec(xy) - y}{x}$

The correct choice is (E).

9. To be concave up the second derivative must be positive. This means that the first derivative must be increasing, which will show up in the tables as an increase in the differences between the values in the second column. This occurs only in table (B).

x	$g(x)$	Difference
1	4	
2	6	2
3	9	3
4	14	5

The correct choice is (B).

10. $\int \dfrac{dx}{x^2+4x} = \int \dfrac{dx}{x(x+4)}$

Decompose $\dfrac{1}{x(x+4)}$ into partial fractions:

$\dfrac{1}{x(x+4)} = \dfrac{A}{x} + \dfrac{B}{x+4} = \dfrac{A(x+4)+Bx}{x(x+4)} = \dfrac{Ax+4A+Bx}{x(x+4)} = \dfrac{x(A+B)+4A}{x(x+4)}$

Since $\dfrac{1}{x(x+4)} = \dfrac{x(A+B)+4A}{x(x+4)}$, equate the coefficients in the numerators and consequently $4A=1$ and $A+B=0 \Rightarrow A=\frac{1}{4}$ and $B=-\frac{1}{4}$.

Therefore, $\int \dfrac{dx}{x^2+4x} = \int \dfrac{dx}{4x} - \int \dfrac{dx}{4(x+4)}$

The correct choice is (E).

11. You are to inscribe 4 rectangles of equal width on $[0,4]$ for the function $f(x) = x^2 - 6x + 9 = (x-3)^2$. Each rectangle will have a width of 1. However, only 2 rectangles can be actually be inscribed; the other 2 have a height of zero. The height of the first rectangle on $[0,1]$ is $f(1)=4$, and the height of the second rectangle on $[1,2]$ is $f(2)=1$. Therefore, the approximate area is $1(4+1) = 5$.

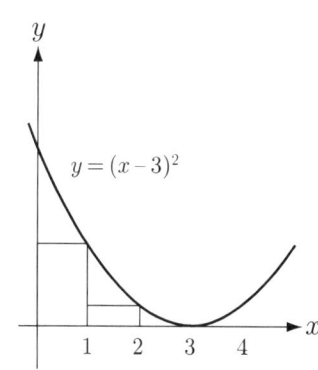

The correct choice is (D).

12. Note the pattern in the derivatives of $y = \sin(2x)$:

$y = \sin(2x)$

$y' = 2\cos(2x) = 2^1 \cos(2x)$

$y'' = -4\sin(2x) = -2^2 \sin(2x)$

$$y''' = -8\cos(2x) = -2^3\cos(2x)$$

$$y^{(4)} = 16\sin(2x) = 2^4\sin(2x)$$

$$y^{(5)} = 32\cos(2x) = 2^5\cos(2x)$$

Each time you take the derivative, the chain rule introduces another factor of 2.

Also note that every "even" derivative of $y = \sin(2x)$ contains $\sin(2x)$ as a factor. So the 20th derivative will contain $\sin(2x)$ as a factor, eliminating choices (C), (D), and (E). The 20th derivative is "similar" to the 4th derivative but the coefficient is 2^{20}.

The correct choice is (B).

13. Begin by finding where $f'(x) = 3$:

$$f'(x) = 7 - 2x$$
$$3 = 7 - 2x$$
$$-4 = -2x$$
$$2 = x$$

Find the point of tangency: $f(2) = 7(2) - 2^2 = 10$. The equation of the line through $(2, 10)$ with a slope of 3 is

$$y - 10 = 3(x - 2)$$
$$y - 10 = 3x - 6$$
$$y = 3x + 4$$

The correct choice is (B).

14. Statement I need not be true. Consider $f(x) = x^3$ on $[-1, 2]$. $f'(0) = 0$, but $f(-1) \neq f(2)$.

Statement II need not be true. Again, consider $f(x) = x^3$ on $[-1, 2]$. $f'(0) = 0$, but $x = 0$ is not a relative extremum (maximum or minimum).

Statement III need not be true. Consider $f(x) = x^3 + 2x^2$. Although $f'(0) = 0$, $f''(0) \neq 0$.

The correct choice is (A).

15. The (negative) velocity decreases quickly and then levels off, its graph is similar to (A). The acceleration is the slope (derivative) of the velocity. So it should be negative, getting less negative and leveling off near zero. Graph (D) is the graph which has these characteristics.

The correct choice is (D).

16. The Taylor series (about $x = 0$) for e^x is $e^x = 1 + x + \dfrac{x^2}{2!} + \dfrac{x^3}{3!} + \cdots$

    Substituting $(-4x)$ for x, we get $e^{-4x} = 1 + (-4x) + \dfrac{(-4x)^2}{2!} + \dfrac{(-4x)^3}{3!} + \cdots$

    So $e^{-4x} \approx 1 - 4x + \dfrac{16x^2}{2} - \dfrac{64x^3}{6} = 1 - 4x + 8x^2 - \dfrac{32}{3}x^3$

    The correct choice is (D).

17. The length of the graph of f from $x = 0$ to $x = 2$ is $L = \displaystyle\int_0^2 \sqrt{1 + (f'(x))^2}\, dx$.

    Since $f'(x) = \sqrt{x^2 - 2x}$ (as given in the problem), $(f'(x))^2 = x^2 - 2x$.

    Therefore, $L = \displaystyle\int_0^2 \sqrt{x^2 - 2x + 1}\, dx$

    $= \displaystyle\int_0^2 |x - 1|\, dx$

    $= \displaystyle\int_0^1 (1 - x)\, dx + \int_1^2 (x - 1)\, dx$

    $= \left(x - \dfrac{x^2}{2}\right)\bigg|_0^1 + \left(\dfrac{x^2}{2} - x\right)\bigg|_1^2$

    $= \left[\left(1 - \dfrac{1}{2}\right) - 0\right] + \left[(2 - 2) - \left(\dfrac{1}{2} - 1\right)\right]$

    $= \left[\left(\dfrac{1}{2}\right) - 0\right] + \left[(0) - \left(-\dfrac{1}{2}\right)\right]$

    $= 1$

    The correct choice is (D).

18. Let $u = \ln x \Rightarrow du = \dfrac{dx}{x}$

    $\displaystyle\int_e^{e^2} \dfrac{dx}{x \ln x} = \int_{x=e}^{x=e^2} \dfrac{du}{u} = \ln |u| = \ln |\ln x|\bigg|_e^{e^2} = \ln(\ln e^2) - \ln(\ln e) = \ln 2 - \ln 1 = \ln 2$

    Note: $\ln e^2 = 2$, $\ln e = 1$, and $\ln 1 = 0$

    The correct choice is (A).

19. Since $g(x) = f(3x)$, using the Chain Rule to differentiate gives $g'(x) = 3f'(3x)$.

    Therefore, $g'(0.1) = 3f'(0.3) = 3(1.096) = 3.288$

    The correct choice is (E).

20. The two circles, $r = 2\cos\theta$ and $r = 2\sin\theta$, intersect at the pole since $(0, \frac{\pi}{2})$ satisfies the first equation and $(0, 0)$ the second equation. The other intersection point $(\sqrt{2}, \frac{\pi}{4})$ is where $2\cos\theta = 2\sin\theta$, or $\sin\theta = \cos\theta$.

Therefore, the area $A = 2\int_0^{\frac{\pi}{4}} \frac{1}{2}(2\sin\theta)^2 \, d\theta = 4\int_0^{\frac{\pi}{4}} \sin^2\theta \, d\theta$

The correct choice is (B).

21. This question is testing whether or not you can recognize the limit as the derivative of the function $f(x) = 2x^5 - 5x^3$.

By definition, $f'(x) = \lim\limits_{h \to 0} \dfrac{f(x+h) - f(x)}{h} = \lim\limits_{h \to 0} \dfrac{(2(x+h)^5 - 5(x+h)^3) - (2x^5 - 5x^3)}{h}$

$= \lim\limits_{h \to 0} \dfrac{2(x+h)^5 - 5(x+h)^3 - 2x^5 + 5x^3}{h}$

Therefore the answer is $f'(x) = 10x^4 - 15x^2$.

The correct choice is (D).

22. Using one of the properties of the definite integral, $\int_2^4 f(x)\,dx + \int_4^8 f(x)\,dx = \int_2^8 f(x)\,dx$.

Or, $6 + \int_4^8 f(x)\,dx = -10 \Rightarrow \int_4^8 f(x)\,dx = -16$.

However, $\int_8^4 f(x)\,dx = -\int_4^8 f(x)\,dx = -(-16)$ or 16.

Note: $\int_a^b f(x)\,dx = -\int_b^a f(x)\,dx$.

The correct choice is (E).

23. $y' = 3x^2 + 2ax + b$

$y'' = 6x + 2a$

Since at $x = 2$ there is a point of inflection, $y''(2) = 0$; therefore, $6(2) + 2a = 0 \Rightarrow a = -6$.

Using $a = -6$ in the original equation and using the fact that when $x = 2$, $y = 0$, $0 = 2^3 - 6(2)^2 + b(2) - 8$.

Therefore solving for b, $b = 12$.

The correct choice is (D).

24. The acceleration vector is the second derivative of the position vector, hence

$$x = 4t^2 \Rightarrow x' = 8t \Rightarrow x'' = 8; \text{ and } y = \sqrt{t} = t^{\frac{1}{2}} \Rightarrow y' = \frac{1}{2}t^{-\frac{1}{2}} \Rightarrow y'' = -\frac{1}{4}t^{-\frac{3}{2}} = \frac{-1}{4t^{\frac{3}{2}}}.$$

Therefore, at $t = 4$, the acceleration vector is $\left(8, -\dfrac{1}{32}\right)$.

The correct choice is (B).

25. Statements I and II are both true by the Extreme Value Theorem.

Statements III is false. Consider $f(x) = x^2$ on $[1, 4]$. $f'(c) \neq 0$ for $1 < c < 4$.

The correct choice is (C).

26. Using the Ratio Test, $\lim\limits_{n \to \infty} \left|\dfrac{u_{n+1}}{u_n}\right| = \lim\limits_{n \to \infty} \left|\dfrac{x^{n+1}}{n+1} \cdot \dfrac{n}{x^n}\right| < 1$

$\Rightarrow \lim\limits_{n \to \infty} \dfrac{n}{n+1}|x| < 1$

$\Rightarrow |x| < 1$ (since $\lim\limits_{n \to \infty} \dfrac{n}{n+1} = 1$)

$\Rightarrow -1 < x < 1$

Now test the endpoints.

When $x = 1$, the series is the alternating harmonic series $1 - \dfrac{1}{2} + \dfrac{1}{3} - \dfrac{1}{4} + \cdots$, which converges.

When $x = -1$, the series is the negative of the harmonic series $-1 - \dfrac{1}{2} - \dfrac{1}{3} - \dfrac{1}{4} + \cdots$, which diverges.

Therefore, the series converges for $-1 < x \leq 1$.

The correct choice is (C).

27. Since $\cos x = \sum\limits_{n=0}^{\infty} \dfrac{(-1)^n x^{2n}}{(2n)!}$, the given series is the Taylor series of $\cos x$ with $x = \pi$.

$\cos \pi = -1$.

The correct choice is (B).

28. Solve the differential equation by separating the variables and integrating both sides of the equation:

$$y\,dy = x\,dx$$

$y^2 = x^2 + C$. Substitute the initial condition $y(3) = 4$ to find C.

$$4^2 = 3^2 + C \Rightarrow C = 7$$

Therefore, $y^2 = x^2 + 7$ or $x^2 - y^2 = -7$

The correct choice is (A).

29. I is the integral which will give the area using vertical rectangles. II and III are identical except for the "dummy" variable of integration; they both represent the area found using horizontal rectangles.

The correct choice is (E).

30. The fourth derivative is part of the coefficient of the fourth-degree term.

$$\frac{f^{(4)}(0)}{4!}x^4 = \frac{x^4}{2(4)}$$

$$f^{(4)}(0) = \frac{4!}{8} = 3$$

The correct choice is (C).

31. An analysis of the derivative, $f'(x)$, may be summarized by the number line

Moving from left to right, f decreases to a minimum at $x = a$, increases to a maximum at $x = b$, decreases to a minimum at $x = c$, and then increases thereafter. These characteristics are shown in graph (B).

The correct choice is (B).

32. This is an initial value problem for which it is difficult or impossible to find an antiderivative. Nevertheless, an expression can be written for the solution. The general form, given $f'(x)$ and $f(c)$, is $f(x) = f(c) + \int_c^x f'(t)\, dt$ by the Fundamental Theorem of Calculus.

Notice: the lower limit of integration is the x-coordinate of the initial condition. The function f starts at this value and accumulates from there for each value of x. Use your calculator to do the computation.

$$f(x) = f(3) + \int_3^x \frac{\sin(1+t^2)}{t^3 - 2t}\, dt$$
$$f(5) = 7 + \int_3^5 \frac{\sin(1+t^2)}{t^3 - 2t}\, dt$$
$$f(5) \approx 6.992$$

The correct choice is (D).

33. The given information implies that the series is absolutely convergent, so II is true.

If a series is absolutely convergent, then it is convergent, so I is true.

By moving the negative sign in front of $\sum$, then III reduces to I, so III is also true.

The correct choice is (E).

34. Graph the function in a suitable window such as $x[-2, 5]$ by $y[-5, 15]$. The graph is tangent to the x-axis at $x = 2$.

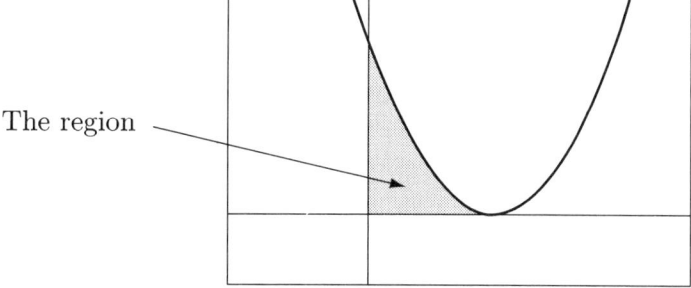

The region

The side of each square cross section lies in the xy-plane and its length is the y-coordinate of the point on the graph; its area is y^2.

The thickness of each piece is in the x-direction and is represented by dx.

The total volume may be found by using the definite integral to sum the individual pieces.

$$\int_0^2 (3(x-2)^2)^2\, dx = 57.6 \text{ on your graphing calculator.}$$

The correct choice is (E).

35. The rate of change of $\sin^2 x$ is given by

$$\frac{d}{dt}(\sin^2 x) = 2\sin x \cos x \frac{dx}{dt}$$

When $x = \frac{\pi}{4}$,

$$\frac{d}{dt}(\sin^2 x)\Big|_{x=\pi/4} = 2\sin\frac{\pi}{4}\cos\frac{\pi}{4}\frac{dx}{dt}$$
$$k\frac{dx}{dt} = 2 \cdot \frac{\sqrt{2}}{2} \cdot \frac{\sqrt{2}}{2}\frac{dx}{dt}$$
$$k\frac{dx}{dt} = \frac{dx}{dt}$$
$$k = 1$$

The correct choice is (C).

36. The Lagrange form of the remainder of a Taylor polynomial is given by

$$R_n(x) = \frac{f^{(n+1)}(Z)}{(n+1)!}(x-c)^{n+1}$$ where Z is some number in the interval of convergence.

Substituting the seventh derivative, $c = 0$ and $x = 1$ into this expression gives

$$R_6(1) = \frac{f^7(Z)}{7!}(1-0)^7 = \frac{10,000\cos(Z)}{7!}(1)$$

Since the largest $\cos(Z)$ can be is $+1$, the largest $R_6(1)$ can be is $\frac{10,000}{7!} = 1.984$.

The correct choice is (E).

37. $f(1) - f(0) = \int_0^1 \frac{\tan^2 x}{x^2+1}\,dx$

$\frac{1}{2} - f(0) \approx 0.3446$

$-f(0) \approx 0.3446 - 0.500$

$f(0) \approx 0.1554$

There is no easy way to find an antiderivative for the given expression. Therefore you must evaluate the definite integral on your calculator.

The correct choice is (B).

38. $f'(0) > 0$ and $f'(6) > 0$ since the function is increasing for $x \leq 8$.

   $f''(4) < 0$ since the graph is concave down for $x \leq 10$.

   $f''(10) = 0$ since this is a point of inflection.

   $f''(12) > 0$ since the graph is concave up for $x \geq 10$.

   Therefore, the smallest value is $f''(4)$.

   The correct choice is (C).

39. Let $p(t) =$ price of the car at time t
    $p'(t) =$ rate of change of the price of the car

   You are given that $p'(t) = 120 + 180\sqrt{t}$.

   You want the price in 5 years, so you need the function $p(t)$.

   $$p(t) = \int p'(t)\, dt = \int (120 + 180\sqrt{t})\, dt = 120t + 120t^{\frac{3}{2}} + C$$

   Since $p(0) = 14,500$, $C = 14,500$.

   Therefore $p(t) = 120t + 120t^{\frac{3}{2}} + 14,500 \Rightarrow p(5) = 120(5) + 120(5)^{1.5} + 14,500 \approx 16,440$.

   This may also be approached as an accumulation function.

   $$p(5) = 14,500 + \int_0^5 (120 + 180\sqrt{x})\, dx \approx 16,440$$

   The correct choice is (C).

40. The area is given by $\int_0^k (2x - \sin x)\, dx = x^2 + \cos x \Big|_0^k = k^2 + \cos k - 1$

   Solve the equation $k^2 + \cos k - 1 = 0.1$ on your calculator; $k = 0.444$.

   The correct choice is (A).

41. Calculate the change in the value of the derivative from the previous row. The values are shown in the following table:

x	$f'(x)$	$\Delta f'(x)$
0.998	0.980	
0.999	0.995	0.015
1.000	1.000	0.005
1.001	0.995	−0.005
1.002	0.980	−0.015

The change in the derivative is first positive then negative, indicating that the slope increases then decreases. This indicates a point of inflection in the range covered by the table.

The correct choice is (C).

42. The fifth week extends from $t = 4$ to $t = 5$. The average rate of change is given by

$$\frac{m(5) - m(4)}{5 - 4} = \frac{9.2758 - 2.7299}{1} = 6.546$$

The values are found by evaluating the function with a calculator.

The correct choice is (D).

43. **Method I:**

The integrals give the area of the region between the x-axis and the graph of the function. Since the region below the x-axis is counted as negative the greatest value will be $\int_0^2 f(x)dx$.

Method II:

The choices may be thought of as various values of the function $F(t) = \int_0^t f(x)\,dx$. Since $F'(t) = f(t)$ by the Fundamental Theorem of Calculus, the largest value of F will occur when its derivative, f, changes from positive to negative. From the graph this occurs only at $t = 2$, so $F(2) = \int_0^2 f(x)\,dx$ will be the absolute maximum.

The correct choice is (B).

44. First find the position equation by integration:

$x'(t) = 3 - 4\cos t$ $\qquad\qquad\qquad$ $y'(t) = 4\sin t$

$x(t) = 3t - 4\sin t + C$ $\qquad\qquad$ $y(t) = -4\cos t + K$

$x(0) = 3(0) - 4\sin 0 + C \Rightarrow C = 0$ $\qquad$ $y(0) = -4\cos 0 + K \Rightarrow -4 + K = -1 \Rightarrow K = 3$

$x(t) = 3t - 4\sin t$ $\qquad\qquad\qquad$ $y(t) = 3 - 4\cos t$

Then graph the position equation in parametric mode in a suitable window such as $x[-2, 40]$ by $y[-2, 8]$. Choice (E) best describes the graph.

The correct choice is (E).

45. The limit is the Riemann sum for the function $x \cos x$ on the closed interval $[0, \pi]$. Therefore,
$$\lim_{n \to \infty} \sum_{i=1}^{n} x_i \cos(x_i) \Delta x = \int_0^\pi x \cos x \, dx = -2,$$ by using your graphing calculator.

The correct choice is (A).

1a. Graph each parametric pair of equations on your calculator and copy the results on the axis provided.

$n = 3$

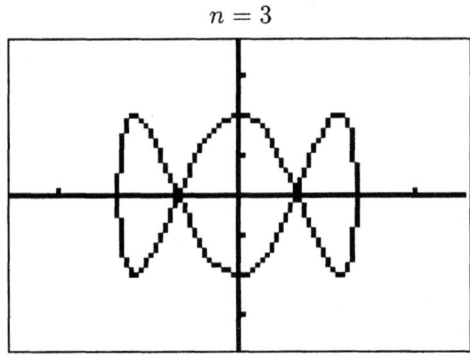

$n = 4$

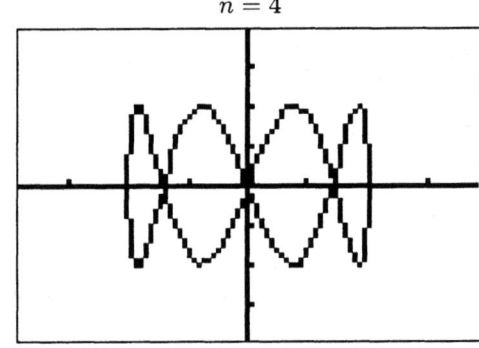

1b. $\dfrac{dy}{dx} = \dfrac{dy/dt}{dx/dt} = \dfrac{n\cos(nt)}{-\sin t}$

1c. There are $2n$ points at which the tangent line will be horizontal. These occur when the derivative is zero. The derivative will be zero only when the factor $\cos(nt) = 0$. Since $0 \le t \le 2\pi$, $0 \le nt \le 2\pi n$. The $\cos(nt)$ will be equal to zero twice in each interval of length 2π. Since there are n intervals of this length between zero and $2\pi n$, the derivative will be zero $2n$ times.

1d. The tangent line will be vertical when the derivative, $\frac{dy}{dx}$, is undefined. This occurs when its denominator is zero: $-\sin t = 0 \Rightarrow$ at $t = 0$, π, and 2π.

2a. $f(0) = 1$ given;

$$f'(0) = (-1)af(0) = -a$$

$$f''(0) = (+1)a^2 f(0) = a^2;\text{ and}$$

$$f'''(0) = (-1)a^3 f(0) = -a^3.$$

Use these values to write the first four terms and identify the pattern for the general term:

$$f(x) = 1 - ax + \frac{a^2 x^2}{2!} - \frac{a^3 x^3}{3!} + \cdots + (-1)^n \frac{a^n x^n}{n!} + \cdots$$

2b. $e^{-x} = \sum_{n=0}^{\infty} (-1)^n \frac{x^n}{n!}$ is a familiar series of the same form as the one found in part (a). The only difference is that part (a) has factors of a^n in each term, so the familiar function must be $f(x) = e^{-ax}$.

2c. The series alternates signs so the absolute value of the error will be less than the first omitted term. Substitute $a = 2$ and $x = 0.2$ and solve.

$$\left| \frac{2^n (0.2)^n}{n!} \right| < 0.001 \text{ by substituting values of } n.$$

When $n = 5$, $\dfrac{2^n (0.2)^n}{n!} = 0.0000853$

Therefore, terms with $n > 4$ may be omitted. Thus five terms are necessary ($n = 0, 1, 2, 3,$ and 4) for the required accuracy.

3a. $s(x)$ is the area from $t = 0$ to $t = x$ of the region bounded by the graph and the x-axis, with the region below the x-axis counted as negative.

3b. The region whose area is $s(4)$ is a trapezoid whose area is $\frac{1}{2}(4)(7+3) = 20$. The graph is the derivative of $s(x)$ by the Fundamental Theorem of Calculus. So $s'(4) = f(4) = 3$.

3c. The graph is the derivative of $s(x)$ by the Fundamental Theorem of Calculus. So $s'(x) = 0$ when $f(x) = 0 \Rightarrow x = 10$.

3d. By the Fundamental Theorem of Calculus the graph is the derivative of $s(x)$. Since the derivative changes from positive to negative at $x = 10$, this is the location of the maximum. The function increases from $x = 0$ to $x = 10$ and then decreases after $x = 10$; therefore the endpoints are not candidates for the maximum value.

4a. The midpoints of the 5 intervals are $t = 0.1, 0.3, 0.5, 0.7$, and 0.9. The midpoint sum is

$$(0.2)(90 + 80 + 80 + 40 + 10) = 0.2(300) = 60$$

4b. The antiderivative of acceleration is the velocity. Thus $f(1)$ is the velocity of the train at the end of the first hour. The units are miles/hour.

4c. $v(t) = \int a(t)\, dt = \int 90\, dt$

$v(t) = 90t + C$

$v(0) = 0 = 90(0) + C$

$0 = C$

$s(t) = \int_0^{0.2} v(t)\, dt = \int_0^{0.2} 90t\, dt = 45t^2 \Big|_0^{0.2} = 1.8$

The train traveled 1.8 miles.

5a.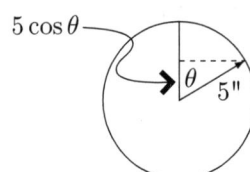

The line indicated in the figure has a length of $5\cos\theta$. Add this to 35 inches that the center is above the floor to get the expression $y = 35 + 5\cos\theta$.

5b. Since the hand travels through 2π radians each 60 minutes, the change in θ is $\frac{d\theta}{dt} = 2\pi \cdot \frac{1}{60} = \frac{\pi}{30}$ radians per minute. By integration, $\theta = \frac{\pi}{30}t$. Since θ and t both start at zero, the constant of integration is zero.

5c. Differentiate your answer to (a) with respect to t and substitute for $\frac{d\theta}{dt}$.

$$\frac{dy}{dt} = -5\sin\theta \cdot \frac{d\theta}{dt} = -5\sin\theta \cdot \frac{\pi}{30} = -\frac{\pi}{6}\sin\theta$$

5d. The answer to (c) gives the rate of change in y. To find when the rate of change is largest (when the increase is most rapid), find its derivative $y'' = -\frac{\pi}{6}\cos\theta\frac{d\theta}{dt}$. Since $\frac{d\theta}{dt}$ is constant, y'' will be zero when $\cos\theta = 0$. This occurs when $\theta = \frac{\pi}{2}$ and $\frac{3\pi}{2}$ or 15 and 45 minutes after the hour. The distance is increasing when the hand is moving upwards; this is at 45 minutes after the hour.

6a. Separate the variables and solve the differential equation:

$$\frac{dT}{dt} = 14 - 0.01T$$

$$\frac{dT}{14 - 0.01T} = dt$$

$$-100 \ln|14 - 0.01T| = t + k_1$$

$$\ln|14 - 0.01T| = -0.01t + k_2, \; k_2 = \frac{k_1}{-100} = -0.01k_1$$

$$14 - 0.01T = k \cdot e^{-0.01t}, \; k = e^{k_2}$$

$$T(t) = -100(k \cdot e^{-0.01t} - 14)$$

$$T(t) = 1400 - k \cdot e^{-0.01t}$$

6b. Substitute $t = 0$ and $T = 0$ to find k:

$$0 = 1400 - k \cdot e^0$$

$$0 = 1400 - k(1)$$

$$k = 1400$$

6c. The temperature is given by $T(t) = 1400 - 1400e^{-0.01t}$.

As $t \to \infty$, that is, if the burner is left on a long time, $\lim_{t \to \infty} T(t) = 1400$.

The hottest the burner will ever get is about $1400°F$ above room temperature. Since this is less that $1500°F$, the burner is safe.

Note: $\lim_{t \to \infty}(1400 - 1400e^{-0.01t}) = \lim_{t \to \infty} 1400(1 - e^{-0.01t}) = 1400$

Sample Examination VI

1. The segments in the slope field will be horizontal when the numerator $x^2y + y^2x = xy(x+y) = 0$. This occurs when $x = 0$, or $y = 0$, or $y = -x$.

   The correct choice is (E).

2. This problem is solved using integration by parts. $u = x, du = dx, dv = e^x\,dx, v = e^x$.

   $$\int_0^2 xe^x\,dx = uv - \int v\,du = xe^x - \int e^x\,dx = xe^x - e^x\Big|_0^2 = (2e^2 - e^2) - (0 - e^0) = e^2 + 1$$

   The correct choice is (C).

3. The y-component of the acceleration is y''.

   $y' = \alpha\beta \cos\beta t$
   $y'' = \alpha\beta(-\beta \sin\beta t)$
   $\quad = -\beta^2 \alpha \sin\beta t$
   $\quad = -\beta^2 y \quad$ (since $y = \alpha \sin\beta t$)

   The correct choice is (A).

4. If $y = \dfrac{3x+4}{4x-3}$, then by the Quotient Rule, $y' = \dfrac{3(4x-3) - 4(3x+4)}{(4x-3)^2}$,

   or $y' = \dfrac{-25}{(4x-3)^2}$. When $x = 1$, $y' = -25$.

   An equation of the tangent line is $y - 7 = -25(x-1)$ or $y = -25x + 32$.

   The correct choice is (A).

5. The acceleration of a particle $a(t) = v'(t)$ where $v(t) = x'(t)$.

   $$x(t) = \frac{1}{2}\sin t + \cos(2t)$$
   $$v(t) = x'(t) = \frac{1}{2}\cos t - 2\sin(2t)$$
   $$a(t) = v'(t) = -\frac{1}{2}\sin t - 4\cos(2t)$$

   Therefore $a(\frac{\pi}{2}) = -\frac{1}{2}\sin(\frac{\pi}{2}) - 4\cos(\pi) = -\frac{1}{2} - 4(-1) = \frac{7}{2}$

   The correct choice is (E).

6. By the Chain Rule $h'(x) = f'(g(x))g'(x)$. Thus
   $h'(-3) = f'(g(-3))g'(-3) = f'(1)g'(-3) = \left(-\frac{1}{3}\right)(-1) = +\frac{1}{3}$

   The value of $g(-3)$ is read from the graph of $g(x)$, and the derivatives are the slopes of the lines at $g(-3)$ and $f(1)$.

   The correct choice is (D).

7. The number may be represented as an infinite geometric series.

   $N = 0.242424\cdots = 0.24 + 0.0024 + 0.000024 + \cdots$
   $= 24\left(\frac{1}{100}\right)^1 + 24\left(\frac{1}{100}\right)^2 + 24\left(\frac{1}{100}\right)^3 + \cdots$

   Therefore $N = \sum_{k=1}^{\infty} 24\left(\frac{1}{100}\right)^k$. So I is true.

   The sum of this geometric series is given by

   $N = \frac{a_1}{1-r} = \frac{24\left(\frac{1}{100}\right)}{1-100^{-1}} = \frac{\frac{24}{100}}{\frac{99}{100}} = \frac{24}{99} = \frac{8}{33}$. So II is false and III is true.

   The correct choice is (E).

8. Let $u = x^2 + 16$, so $du = 2x\, dx \Rightarrow x\, dx = \frac{1}{2} du$

   $\int \frac{x}{\sqrt{x^2+16}}\, dx = \frac{1}{2}\int \frac{du}{\sqrt{u}} = \frac{1}{2}\int u^{-\frac{1}{2}}\, du = \frac{1}{2}(2u^{\frac{1}{2}}) = u^{\frac{1}{2}} = (x^2+16)^{\frac{1}{2}} = \sqrt{x^2+16}$

   Therefore, $\int_0^3 \frac{x}{\sqrt{x^2+16}}\, dx = \sqrt{x^2+16}\Big|_0^3 = \sqrt{25} - \sqrt{16} = 5 - 4 = 1$.

   The correct choice is (A).

9. $y = \ln(3x+5)$

   $\frac{dy}{dx} = \frac{3}{3x+5}$ $\left(\text{The derivative of } \ln u \text{ is } \frac{du}{u}.\right)$

   $\frac{d^2y}{dx^2} = \frac{-9}{(3x+5)^2}$

   The correct choice is (D).

10. Using the method of partial fractions,

$$\frac{1}{x^2+5x+6} = \frac{1}{(x+2)(x+3)} = \frac{A}{x+2} + \frac{B}{x+3}$$
$$1 = A(x+3) + B(x+2)$$

Let $x = -3$. Then $1 = -B \Rightarrow B = -1$

Let $x = -2$. Then $1 = A \Rightarrow A = 1$

$$\int \frac{dx}{x^2+5x+6} = \int \frac{dx}{x+2} - \int \frac{dx}{x+3}$$
$$= \ln|x+2| - \ln|x+3| = \ln\left|\frac{x+2}{x+3}\right|$$

Therefore, $\ln\left|\frac{x+2}{x+3}\right|\Big|_{-1}^{1} = \ln\left(\frac{3}{4}\right) - \ln\left(\frac{1}{2}\right)$
$$= (\ln 3 - \ln 4) - (\ln 1 - \ln 2)$$
$$= \ln 3 - 2\ln 2 - \ln 1 + \ln 2$$
$$= \ln 3 - \ln 2$$
$$= \ln\left(\frac{3}{2}\right)$$

The correct choice is (A).

11. By implicit differentiation, $2yy' - [(2x)(y') + 2y] = 0$
$$2yy' - 2xy' - 2y = 0$$
$$2yy' - 2xy' = 2y$$
$$y'(2y - 2x) = 2y$$
$$y' = \frac{2y}{2y-2x} = \frac{2y}{2(y-x)}$$
$$y' = \frac{y}{y-x}\Big|_{(2,-3)} = \frac{-3}{-3-2} = \frac{-3}{-5} = \frac{3}{5}$$

The correct choice is (E).

12. The average value of $\sqrt{3x}$ on $[0,9]$ is $\frac{1}{9-0}\int_0^9 \sqrt{3x}\, dx$.

$$\int \sqrt{3x}\, dx = \sqrt{3}\int \sqrt{x}\, dx = \frac{2\sqrt{3}}{3}x^{\frac{3}{2}}.$$

Therefore, the average value is $\frac{1}{9}\left(\frac{2\sqrt{3}}{3}x^{\frac{3}{2}}\right)\Big|_0^9 = 2\sqrt{3}$.

The correct choice is (B).

13. The derivative represents the instantaneous change in C with respect to x; thus its units are the units of C divided by the units of x: dollars per item. I is true.

The derivative represents the additional cost of producing one more item (i.e. the change in cost per item produced). II is true.

The units for the rate at which items are produced would be items per time. III is false.

The correct choice is (D).

14. $\int_{-\infty}^{\infty} e^{-|x|} \, dx = \int_{-\infty}^{0} e^{x} \, dx + \int_{0}^{\infty} e^{-x} \, dx$

$\int_{-\infty}^{0} e^{x} \, dx = \lim_{t \to -\infty} (e^{x}) \Big|_{t}^{0} = \lim_{t \to -\infty} (1 - e^{t}) = 1$

$\int_{0}^{\infty} e^{-x} \, dx = \lim_{t \to \infty} (-e^{-x}) \Big|_{0}^{t} = \lim_{t \to \infty} (1 - e^{-t}) = 1$

Therefore, $\int_{-\infty}^{\infty} e^{-|x|} \, dx = 1 + 1$ or 2.

The correct choice is (D).

15. $y = 6x^2 + \frac{x}{2} + 3 + 6x^{-1}$

$y' = 12x + \frac{1}{2} - 6x^{-2}$

$y'' = 12 + 12x^{-3}$

$y'' = 0$ when $12 + 12x^{-3} = 0$ or $12(1 + x^{-3}) = 0 \Rightarrow x = -1$

Note: y'' does not exist at $x = 0$.

Signs of $\frac{y''}{x}$: $+$ on $(-\infty, -1)$, $-$ on $(-1, 0)$, $+$ on $(0, \infty)$

Therefore, the graph of y is concave down for $-1 < x < 0$ since $y'' < 0$.

The correct choice is (C).

16. Let (x, y) be a point on the parabola $y = 12 - x^2$.

Then the area of the rectangle, $A = (2x)(12 - x^2) = 24x - 2x^3$.

$A' = 24 - 6x^2 = -6(x^2 - 4) = -6(x - 2)(x + 2) \Rightarrow A' = 0$ when $x = \pm 2$.

The maximum occurs when $x = 2$. So the area of the rectangle is $A(2) = 32$.

The correct choice is (D).

17. The units of an accumulation function are the product of the units of the dependent and independent variables. Thus

$$\frac{liters}{kilometer} \cdot kilometers = liters$$

If the function $r(c)$ were graphed, the definite integral would give the area under the graph. The units of the horizontal dimension of the "squares" are those of the independent variable, kilometers. The units of the vertical dimension are those of the dependent variable, liters/kilometer. The units of the area of each square is the product of the horizontal and vertical dimensions: liters.

The correct choice is (A).

18. The graph of $f(x)$ has 2 relative maxima and 2 relative minima, which means that $f'(x)$ crosses the x-axis in four points.

The correct choice is (E).

19. Use implicit differentiation to find the derivative:

$$2x - 2y\frac{dy}{dx} + 1 = 0$$

$$\frac{dy}{dx} = \frac{2x + 1}{2y}$$

The derivative is undefined and the tangent is vertical when $y = 0$. When $y = 0$

$$x^2 - 0 + x = 2$$
$$x^2 + x - 2 = 0$$
$$(x + 2)(x - 1) = 0$$
$$x = -2, 1$$

So the points are $(-2, 0)$ and $(1, 0)$.

The correct choice is (D).

20. The area enclosed by the graph and the t-axis is the distance the object has moved from the starting point. When the velocity is positive (above the t-axis) the object is moving to the right; when the velocity is negative (below the t-axis) the object is moving to the left. It moves to the right from the beginning, past point A and point B to point C when it begins moving to the left. Therefore, it is farthest to the right at point C.

    The correct choice is (C).

21. The definition of the derivative of f at $x = 2$ is: $f'(2) = \lim\limits_{x \to 2} \dfrac{f(x) - f(2)}{x - 2}$.

    Therefore, from the given information, $\lim\limits_{x \to 2} \dfrac{f(x)}{x - 2} = \lim\limits_{x \to 2} \dfrac{f(x) - f(2)}{x - 2} \Rightarrow f(2) = 0$, so I is true.

    Since $f'(2) = 0$ and $f(2) = 0$, the graph of $f(x)$ has a horizontal tangent at the point $(2, 0)$, so III is true.

    If a function is differentiable at a point, it must be continuous there. Therefore, since f is differentiable at $x = 2$ $(f'(2) = 0)$, f is continuous at $x = 2$, so II is true.

    The correct choice is (E).

22. The antiderivative is found by using the method of Integration by Parts.

    Let $u = x, du = dx, dv = e^{-x} dx$ and $v = -e^{-x}$. Then

    $$\int xe^{-x}\, dx = -xe^{-x} - \int (-e^{-x})\, dx = -xe^{-x} - e^{-x} + C$$

    $$\int_1^\infty xe^{-x}\, dx = \lim_{b \to \infty} \int_1^b xe^{-x}\, dx = \lim_{b \to \infty}\left(-xe^{-x} - e^{-x}\Big|_1^b\right) = \lim_{x \to \infty}(-be^{-b} - e^{-b} + e^{-1} + e^{-1}) = 2e^{-1} = \dfrac{2}{e}$$

    The correct choice is (C).

23. Since $\sin^2 x = 1 - \cos^2 x = (1 - \cos x)(1 + \cos x)$,

    $$f(x) = \dfrac{\sin^2 x}{1 - \cos x} = \dfrac{(1 - \cos x)(1 + \cos x)}{1 - \cos x} = 1 + \cos x$$

    Therefore, $f'(x) = -\sin x$.

    The correct choice is (C).

24. $\frac{dy}{dx} = y\cos x \Rightarrow \frac{dy}{y} = \cos x \, dx$

$\Rightarrow \int \frac{dy}{y} = \int \cos x \, dx$

$\Rightarrow \ln|y| = \sin x + C$

$\Rightarrow |y| = e^{\sin x + C} = e^C e^{\sin x}$

$\Rightarrow y = C_1 e^{\sin x}$

Since $y = 3$ when $x = 0$, $3 = C_1 e^{\sin x}$

$3 = C_1(1)$

Therefore, $y = 3e^{\sin x}$

The correct choice is (E).

25. The fact that the definite integral is larger than T indicates that the slanted sides of the trapezoids all lie below the graph of $f(x)$ (even if $f(x) < 0$). Thus, the curve must be concave downwards.

The function $f(x)$ cannot be linear or the definite integral would equal T. Increasing or decreasing makes no difference; the function could be either.

The correct choice is (C).

26. $f'(x) = 2(x+2)^3(x-5) + 3(x+2)^2(x-5)^2$
$= (x+2)^2(x-5)(2(x+2) + 3(x-5))$
$= (x+2)^2(x-5)(5x-11)$

Since $(x+2)^2 \geq 0$, the derivative changes sign only twice. Therefore, there are only two extreme values.

The correct choice is (C).

27. The limit may be treated as a Riemann sum for some function on an interval. The general form of a Riemann sum is $\sum_{k=1}^{n} f(x_k)(\frac{b-a}{n})$ where the x_k are values in each subinterval. If $b - a = 3$ then the entire interval is three units wide. The first point in the interval occurs when $k = 1$ and $x_1 = 2 + \frac{3}{n}$ so this is a right hand sum; the final value is $x_n = 2 + \frac{3}{n}n = 5$ and the function is $f(x) = x^2$.

The limit is then equal to $\int_2^5 x^2 \, dx = \frac{x^3}{3}\Big|_2^5 = \frac{1}{3}(5^3 - 2^3) = 39$.

Note: There are always more than one solution to this type problem, all of which give the same answer. Another choice is $\int_0^3 (2+x)^2 \, dx = 39$

The correct choice is (D).

28. Statement I is true – Euler's method uses the point on the tangent line with the same x-coordinate to approximate the value of a function.

Statements II and III are false – Euler's method is not used for finding roots. While Euler's method uses the y-coordinate of B to approximate the y-coordinate of D, there is no way of knowing in advance that the y-coordinate of D is zero.

The correct choice is (A).

29. Let $u = 2x$, so $du = 2 \, dx \Rightarrow dx = \frac{1}{2} \, du$, when $x = 2$, $u = 4$ and when $x = 3$, $u = 6$

$$\int_2^3 f(2x) \, dx = \frac{1}{2}\int_4^6 f(u) \, du \Rightarrow \int_4^6 f(u) \, du = 2\int_2^3 f(2x) \, dx$$

Therefore, $\int_4^6 f(x) \, dx = 2(8) = 16$

The correct choice is (D).

30. Area, $A = \frac{1}{2}\int_0^\pi (6\cos\theta + 8\sin\theta)^2 d\theta = 25\pi \approx 78.540$.

This could be done by paper and pencil, but the better approach is with the use of a calculator. The equation is a circle with center at $(3, 4)$ and radius 5.

The correct choice is (C).

31. The volume may be found by multiplying the cross section area, $\int_{-20}^{20} \left(20 - \frac{x^6}{3,200,000}\right) dx$ square feet, by the base (50 feet). It may also be found by forming a Riemann sum of the volumes of slices perpendicular to the x-axis. The base is 50, the height is $y(x)$ and the thickness is dx.

In both cases the volume is given by $50 \int_{-20}^{20} \left(20 - \frac{x^6}{3,200,000}\right) dx \approx 34,300$ cubic feet.

The correct choice is (D).

32. Find where the walls hit the ground by solving

$$20 - \frac{x^6}{3,200,000} = 0$$
$$x^6 = 64,000,000$$
$$x = \pm 20$$

Then the average height is given by $\frac{1}{20-(-20)} \int_{-20}^{20} \left(20 - \frac{x^6}{3,200,000}\right) dx \approx 17.143$ feet, by calculator.

The correct choice is (E).

33. The profit function is $P(x) = R(x) - C(x)$, where

$R(x)$ = revenue (sales) function

$C(x)$ = cost function

If x calculators are sold and the price per calculator is $75 - 0.01x$ dollars each,

then $R(x) = x(75 - 0.01x)$ and $C(x) = 1850 + 28x - x^2 + 0.001x^3$.

The profit function, $P(x) = x(75 - 0.01x) - (1850 + 28x - x^2 + 0.001x^3)$
$= -0.001x^3 + .99x^2 + 47x - 1850$

To maximize $P(x)$, find $P'(x) = -0.003x^2 + 1.98x + 47$ and set $P'(x)$ equal to zero.

Solving this equation on your calculator, $x \approx 683$.

Note: Since $P''(x) = -0.006x + 1.98$ and $P''(683) < 0 \Rightarrow x = 683$ is a maximum.

The correct choice is (C).

34. The area of the shaded region is bounded by the x-axis and $y = f(x)$.

Since $y = 0 \geq y = f(x)$ on $[a, b]$, the area is

$$\int_a^b [0 - f(x)]\, dx = \int_a^b -f(x)\, dx = -\int_a^b f(x)\, dx \text{ or } \int_b^a f(x)\, dx.$$

Keep in mind, that $\int_a^b f(x)\, dx = -\int_b^a f(x)\, dx$.

The correct choice is (B).

35. Referring to the accompanying diagram,

$$\tan \theta = \frac{x}{75,000} \Rightarrow \theta = \text{Arctan}\left(\frac{x}{75,000}\right).$$

Therefore, $\dfrac{d\theta}{dt} = \dfrac{(1/75,000)(dx/dt)}{1 + (x/75,000)^2}$

Since $\dfrac{dx}{dt} = 16,500$ when $x = 38,000$

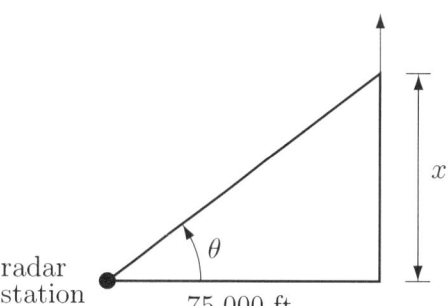

$$\frac{d\theta}{dt} = \frac{16,500/75,000}{1 + (38,000/75,000)^2} = \frac{165/750}{1 + (38/75)^2} \approx 0.175$$

The correct choice is (A).

36. The error in a Taylor Series is given by the expression $\left|\dfrac{f^{(n+1)}(z)}{(n+1)!}(x-a)^{n+1}\right|$ where z is a number between x and a. For the numbers given the derivative is always less than one; then with $x = 1.2$ the error is always less than

$$\left|\frac{1}{(n+1)!}(1.2 - 1)^{n+1}\right|$$

Evaluate this expression on your calculator for the answer choices to determine which is less than 0.00001.

For $n = 3$, the value is 0.0000667; and for $n = 4$, the value is 0.00000267. Therefore, five terms (those with $n = 0, 1, 2, 3$ and 4) are sufficient to approximate the function to the desired accuracy.

The correct choice is (C).

37. By the Integral Test $\int_1^\infty \frac{1}{x^{\pi p}}\, dx$ will converge if the series $\sum_{n=1}^\infty \frac{1}{n^{\pi p}}$ converges. This series is a p-series and will converge if the exponent $\pi p > 1$. Thus, the series and the integral will converge when $p > \frac{1}{\pi}$.

The correct choice is (B).

38. The graph of a function will be concave up when $f''(x) > 0$. Since $f''(x) > 0$ for $x < 0$ and $b < x < c$, I is true.

The (relative) maximum and minimum values of a function cannot be determined from the second derivative alone; II is false.

The function f has points of inflection where $f''(x)$ changes sign. This occurs at $x = 0$ and $x = b$. Thus $x = 0$ and $x = b$ are points of inflection; III is true.

The correct choice is (D).

39. The formula for the length of $g(x)$ is: $L = \int_a^b \sqrt{\left(\frac{dg}{dx}\right)^2 + 1}\, dx$.

By the Fundamental Theorem of Calculus $\frac{dg}{dx} = \sqrt{(f(x))^2 - 1}$ and $\left(\frac{dg}{dx}\right)^2 = (f(x))^2 - 1$.

Substituting this into the first integral gives

$$\int_a^b \sqrt{\left(\frac{dg}{dx}\right)^2 + 1}\, dx = \int_a^b \sqrt{(f(x))^2 - 1 + 1}\, dx$$
$$= \int_a^b f(x)\, dx$$

The correct choice is (A).

40. The term with $k = 10$ of any Taylor series is $f^{(10)}(a)\frac{(x-a)^{10}}{10!}$.

For this series, the corresponding term is $\left(\frac{1}{2}\right)^{10}\frac{(x-3)^{10}}{10!}$.

Therefore, $f^{(10)}(3) = \left(\frac{1}{2}\right)^{10} = 9.766 \times 10^{-4}$.

The correct choice is (C).

41. By the Fundamental Theorem of Calculus,

$$\frac{d}{dx}\int_0^{2x}(e^t+2t)\,dt = 2(e^{2x}+2(2x)) = 2e^{2x}+8x.$$

The correct choice is (E).

42. By the Mean Value Theorem
$$f'(c) = \frac{f(b)-f(0)}{b-0}$$
$$\cos(4.522) = \frac{\sin b}{b}$$

Solving this equation with a calculator gives $b \approx 4$.

The correct choice is (E).

43. The fourth-degree Taylor Polynomial for $\cos(x) = 1 - \frac{x^2}{2!} + \frac{x^4}{4!}$.

Substituting $x = \frac{1}{2}$ gives $1 - \frac{\left(\frac{1}{2}\right)^2}{2!} + \frac{\left(\frac{1}{2}\right)^4}{4!} = 1 - \frac{1}{8} + \frac{1}{384}$

The correct choice is (E).

44. $\int_0^{1000} 8^x\,dx - \int_a^{1000} 8^x\,dx = \int_0^{1000} 8^x\,dx + \int_{1000}^a 8^x\,dx = \int_0^a 8^x\,dx = \left.\frac{8^x}{\ln 8}\right|_0^a = \frac{8^a-1}{\ln 8}$

Solving $\frac{8^a-1}{\ln 8} = 10.40$ on your calculator results in $a = 1.5$.

The correct choice is (B).

45. The average speed is given by $\frac{1}{32,000}\int_0^{32,000} s(a)\,da$. While it is possible to write a piecewise function for $s(a)$ it is easier to find the value of the integral by find the area under the graph directly.

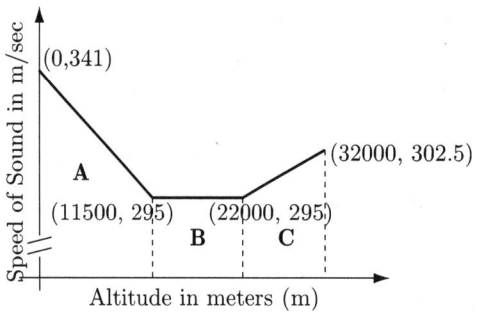

Region A is a trapezoid. The area is $\frac{1}{2}(11,500)(341+295) = 3,657,000$

Region B is a rectangle. The area is $295(10,500) = 3,097,500$

Region C is a trapezoid. The area is $\frac{1}{2}(10,000)(302.5+295) = 2,987,500$

The total area is 9,742,000.

Divide by 32,000 to find the average value = 304.4 m/sec.

Note: It is incorrect to find the average speed on each part and then average the averages. This gives choice (B) 303.9.

The correct choice is (C).

1a.

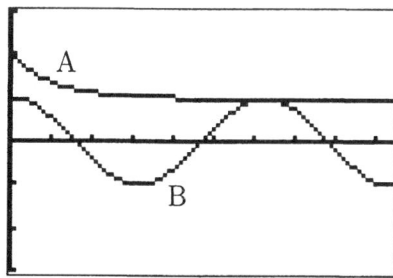

1b. No, since for all x, $1 + e^{-x} > 1$ and $\cos x \leq 1$, the particles can never be at a point with the same y-coordinate, therefore they will never collide.

1c. The distance between the particles is $s(x) = 1 + e^{-x} - \cos x$. Differentiate to find the critical values:

$$\frac{ds}{dx} = -e^{(-x)} + \sin x$$

Set the derivative equal to zero and solve on your calculator. $x = 0.58853\ldots$, $x = 3.09636\ldots$, and $x = 6.2850\ldots$

Then $s(0.58853) = 0.72338$, $s(3.09636) = 2.04419$ and $s(6.285) = 0.00186$.

To be safe also check the endpoints, $s(0) = 1$ and $s(3\pi) = 2.00008$.

The minimum distance is about 0.00186 or 0.001 or 0.002.

1d. From part (c) the maximum distance is 2.04419 or 2.044

2a. The acceleration is negative when the velocity is decreasing. This occurs during the interval $15 \leq t \leq 45$ seconds.

2b. The average acceleration is the change in the velocity divided by the change in time:
$$\frac{3.1 - 0}{15 - 0} = 0.207 \text{ feet/second/second}$$

2c. Using the left hand rectangles, the right hand rectangles or the Trapezoidal Rule all give the same value: The left hand sum is
$$\frac{30 - 0}{6}(0 + 1.6 + 2.7 + 3.1 + 2.7 + 1.6) = \frac{30}{6}(11.7) = 58.5 \text{ feet}$$

2d. The Ferris Wheel is approximately 58.5 feet in diameter. The integral evaluated in part (c) is an approximation of how far *up* the rider traveled in the first half of one complete revolution of the wheel. This is the diameter.

3a. Since the line touches the y-axis and contains the point $(2,0)$ it must have a negative slope. Since the line always lies above the parabola, the largest slope (least steep negative slope) will occur when the line is tangent to the parabola at $(2,0)$. The slope of the parabola is $\dfrac{df}{dx} = -2x$. At the point $(2,0)$ the slope is -4. Therefore $m \leq -4$.

3b. The line from $(0, 12)$ to $(2, 0)$ has a slope of -6 so
$$A = \int_0^2 \left(-6(x-2) - (4-x^2) \right) \, dx = \frac{20}{3}$$

3c. <u>Method 1</u>:

$$A(m) = \int_0^2 \left(m(x-2) - (4-x^2) \right) \, dx$$
$$A(m) = \left. \frac{m}{2}x^2 - 2mx - 4x + \frac{1}{3}x^3 \right|_0^2$$
$$A(x) = 2m - 4m - 8 + \frac{8}{3}$$
$$A(m) = -2m - \frac{16}{3}$$

<u>Method 2</u>:

The area under the parabola is constant and equal to $\int_0^2 (4-x^2) \, dx = \dfrac{16}{3}$. The region formed by the axes and the line is a triangle with base of 2 and altitude of $-2m$. Thus its area is $\frac{1}{2}(-2m)(2) = -2m$. Subtracting these gives the area of R: $A(m) = -2m - \dfrac{16}{3}$

3d. Differentiate the area with respect to time:
$$\frac{dA}{dt} = -2\frac{dm}{dt}$$
$$\frac{dA}{dt} = -2(-2) = 4$$

Thus the area of R is increasing at the constant rate of 4 square units per second.

4a. Integrate the given equations:

$$x(t) = 4\sin t + C$$

$$y(t) = -\cos t + K$$

The constants C and K are found by using the fact that at $t = 0$ both $x = 0$ and $y = 0$.

So $x(0) = 4\sin 0 + C = 0 \Rightarrow C = 0$

and $y(0) = -\cos 0 + K = 0 \Rightarrow K = 1$.

Therefore, $x(t) = 4\sin t$ and $y(t) = -\cos t + 1$.

4b. When $t = 0$, the particle is at $(0,0)$, and at $t = \frac{\pi}{2}$, the particle is at $(4,1)$.

The average rate of change of y with respect to x is $\frac{\Delta y}{\Delta x} = \frac{1-0}{4-0} = \frac{1}{4}$

4c. Set the instantaneous rate of change, $\frac{dy}{dx} = \frac{dy/dt}{dx/dt} = \frac{\sin(t)}{4\cos(t)} = \frac{1}{4}\tan(t)$, equal to the average rate of change found in part (b) and solve for t.

$$\frac{1}{4}\tan t = \frac{1}{4}$$

$$\tan t = 1 \Rightarrow t = \frac{\pi}{4}$$

5a. Since $\frac{d}{dx}\tan(x) = \sec^2(x)$ the polynomial may be found by differentiating $M(x)$:

$$\frac{d}{dx}M(x) = 1 + x^2 + \frac{2}{3}x^4$$

5b. The 4th degree Maclaurin polynomial for $\frac{\tan(x)}{x}$ is $\frac{x + \frac{1}{3}x^3 + \frac{2}{15}x^5}{x} = 1 + \frac{1}{3}x^2 + \frac{2}{15}x^4$

5c. $\lim\limits_{x \to 0} \frac{\tan(x)}{x} = \lim\limits_{x \to 0} \left(1 + \frac{1}{3}x^2 + \frac{2}{15}x^4\right) = 1$

5d. The answer is exact.

The Maclaurin polynomial found in (b) is equal to the function it approximates at the center of the interval of convergence (in this case at $x = 0$). Even though the other terms of the power series are not known, they all will contain powers of x. These other terms will all approach zero at $x = 0$ and will not affect the limit.

6a.

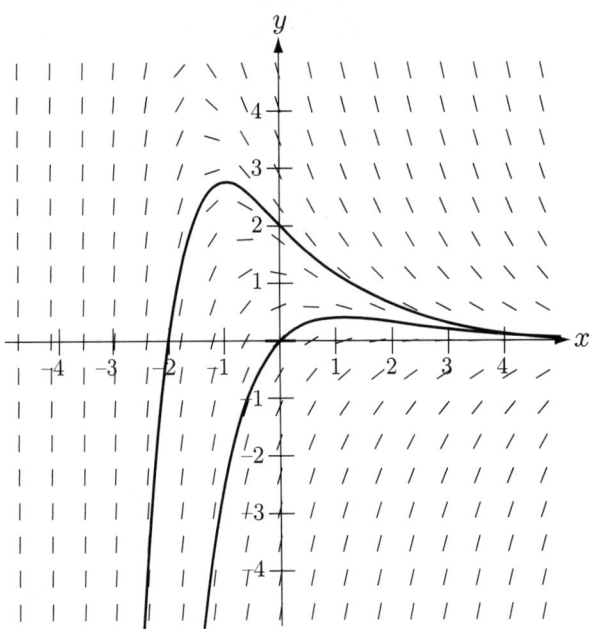

6b.
$$y = xe^{-x} + Ce^{-x}$$
$$\frac{dy}{dx} = -xe^{-x} + e^{-x} - Ce^{-x}$$
$$y + \frac{dy}{dx} = \left(xe^{-x} + Ce^{-x}\right) + \left(-xe^{-x} + e^{-x} - Ce^{-x}\right)$$
$$y + \frac{dy}{dx} = e^{-x}$$

6c.
$$\frac{dy}{dx} = e^{-x}(-x + 1 - C)$$
$$\frac{dy}{dx} = 0 \text{ when } x = 1 - C$$

Substituting $(1-C)$ for x in the solution equation gives
$$y = (1-C)e^{-(1-C)} + Ce^{-(1-C)} = e^{-(1-C)}$$

The maximum points are $\left(1-C, e^{-(1-C)}\right)$. They all lie on the graph of $f(x) = e^{-x}$.

NOTES

NOTES